Midwest Gem, Fossil and Mineral Trails
Prairie States

South Dakota, North Dakota, Nebraska, Iowa, Kansas, Missouri

by
June Culp Zeitner

$7.95

Gem Guides Book Company
3677 San Gabriel Parkway
Pico Rivera, California 90660

GEM GUIDES BOOK CO.

Library of Congress Catalogue Card Number 88-082010
ISBN 935182-40-3

Maps: *JEAN HAMMOND*

Note: Due to the possibility of misinterpretation of information, *Midwest Gem, Fossil and Mineral Trails-The Prairie States,* its author, publisher and all other persons directly or indirectly involved with this publication assume no responsibility for accidents, injury or any losses by individuals or groups using this publication.

In rough terrain and hazardous areas all persons are advised to be aware of possible changes due to man or nature that can occur along the gem trails.

TABLE OF CONTENTS

How to Collect 6
How to Use this Book 7
Region Covered 8
How to Hunt Minerals, Gems and Fossils in the Prairie States 9
The American Federation Code of Ethics 13
South Dakota 15
- Western South Dakota 15
- The Black Hills 25
- Hints for Collecting Minerals 33
- Fossils 35
- Eastern South Dakota 37

North Dakota 41
- Western and Central North Dakota 41
- Hints for Collecting Gem Materials 44
- Eastern North Dakota 48

Nebraska 51
- Western Nebraska 51
- Eastern Nebraska 58
- Hints for Collecting Invertebrate Fossils 62

Iowa 65
- Northern and Central Iowa 65
- Southern Iowa 70

Kansas 81
- Western Kansas 81
- Eastern Kansas 84

Missouri 92
- Western and Central Missouri 93
- Eastern Missouri 95

Bibliography 107
Recommended Additional Reading 108
Some Helps for Field Identification 109

FOREWORD

Midwest Gem Trails covered such an immense and varied territory, that when I was asked to revise it by Al Mayerski, publisher of Gem Guides Books, I suggested that it might be more useful in two volumes, rather than one. Then it was a problem of how best to divide them, since most of the states have something in common geologically with all of their neighbors. For example, the notable geode area of Iowa extends into Illinois and Missouri. The pipestone area of Minnesota extends into South Dakota. There were 12 states, and somehow I had to separate them into two 6s. The two natural divisions are the Great Lakes and the Mississippi River, so this book covers the states of the Midwest which are west of the Mississippi. They have been called the Plains States, the Midwest, the Central States, some of them North Central and some South central, and historically the Prairie States, although Missouri is only partially prairie. However that is the name I like best for this exceptional area for gems, minerals and fossils. The other 6 states of *Midwest Gem Trails* are covered in the Great Lakes States.

There are numerous sources of information for collectors, but this book endeavors to tell not only where to collect and what to collect, but how to prospect on your own and where to get more information and more help. Many of the places in this book have never been described before in any other Guide Book and many of the discoveries I describe are recent, and not well-known outside of the immediate area of the finds.

I have also tried to tell you how to collect safely as well as successfully. You will find many hints on how to make your trips a pleasure. It is my hope this field guide will help you decide to visit some of these sites, and that when you do, you will find your own precious souvenirs from among the unlimited resources of these states.

ACKNOWLEDGEMENTS

Although I have personally visited all of these states many times and have collected at many of the major sites, it is impossible for one person to keep up with the constantly changing conditions, federal or state regulations, changes of ownerships of land, new construction, recent discoveries and more. This is a large area with plenty of opportunities for enthusiastic field collectors. I treasure the beautiful specimens I have found, but even more I treasure the friendships of the warm, generous and wonderful rock hobbyists I have met in these states. I am grateful for the many helps they have given me in preparing this book. Hobbyists have helped me in more ways than I can count, without even knowing they were helping me. An example that comes to mind are the displays at shows of exciting examples of well-labeled local materials.

I am particularly indebted to the following: Willis Smith, Hazel Bishop, Roger Pabian, The Black Hills Institute of Geological Research, Earle and Blossomae Campbell, Gen Burish, Bob Randall, Bob and Maxine Wilson, Jan Baumeister, the Western Dakota Gem and Mineral Society and to two recently deceased dear friends Willard Roberts and Jean Dahlberg. They have all contributed much to the hobby in these states.

Thanks, too, to the museums, Geological Surveys, Tourism Bureaus and rock and mineral dealers who work to increase our knowledge and enjoyment of these mineral, gem, and fossils resources.

Special thanks to my husband Albert, who introduced me to the world of rocks, who took me on over a million miles of rock hunts, and who has been patient and helpful when I was glued to my typewriter.

May my readers all have the pleasures from this hobby that we have had.

Remarkable discoveries are still being made in the late 20th Century. These Columbian mammoth bones, the remains of at least 40 individuals, were discovered a few years ago near Hot Springs, South Dakota.

THE PRAIRIE STATES

HOW TO COLLECT

There have been many changes since the last revision of *Midwest Gem Trails*, and countless changes since I first wrote the book in 1955. Most of the changes have worked against the collector, however much more is known now by amateurs about collecting areas in the 12 states first covered. Professionals are sharing information with amateurs and amateurs are eager to help each other. Every year there are new hobbyists and every year new discoveries are made.

The Prairie States of the Midwest, those which are west of the Mississippi River, east of the Rockies and north of Oklahoma and Arkansas, are an important part of the country for field collectors. The diversity is astounding, ranging from areas rich in invertebrate marine fossils to dinosaur fossils, from crystals in quarries to gold in streams, from sand dunes to igneous mountains. It is a remarkable land with unlimited resources.

What are the changes which have hurt collectors? Land has been withdrawn from Public use by the government. Access to quarries has been severely limited by Federal Mine Safety and Health Rules, and by insurance restrictions. Much gravel is crushed before it is even brought to the surface. Old dump areas are being recovered by vegetation. The doldrums in the mining industry has shut down many mines and quarries. Superhighways and housing developments have covered former collecting sites. Large multi-national corporations are buying up small locally-owned quarries, so the management no longer has an interest in helping local collectors. And in some cases vandalism has forced the closing of productive areas.

Important specimens remain to be found, but it takes a lot more determination and patience than it used to. Our hobby is fairly new as hobbies go, and in reality has merely scratched the surface. In recent years the largest petrified tree of this whole area was found in North Dakota, dinosaurs were discovered in South Dakota as well as the world's largest mammoth site, vivid fortification agates were discovered in Missouri, and peridotites were examined in Kansas for signs of diamond pipes. The six state area has vast resources which will continue to be developed far into the future. Our problems as dedicated field collectors are to try to reopen closed localities, to keep current localities open, and to learn enough about geology and mineralogy that we can discover and develop new localities.

Guided by the AFMS Code of Ethics as well as the Golden Rule, it is up to every rock, gem, mineral, and fossil collector to behave in such a way that we will always be welcomed.

HOW TO USE THIS BOOK

This book is a guide to some excellent collecting areas, but it cannot guarantee that when you get to the area you will be successful. It is up to the individual to get proper permission and to learn enough about geology to find alternative sites. The collector should know how to collect, and should be selective in his collecting.

Although many specific directions are given, you may have better luck with the general information, since places where exact directions are given tend to be worked over in a hurry. The informed collector should read books on mineralogy, paleontology and geology and keep up with magazines in the field. Collectors should exchange information.

The expert collector always goes on a trip well-prepared and willing to work a little harder, dig a little deeper, and examine materials more closely than the novice.

Any collector going on a field trip far from home should make every effort to get all the possible information before making that trip. Land status is ever changing. The availability of materials is not constant. By phone or by letter, one can gather plenty of information. Following are some of the ways to make your trip successful.

1. Contact the rock clubs nearest your destination.
2. Write to Chambers of Commerce, the State Geological Surveys, the State Divisions of Tourism, to museums, mining companies, fee basis operators, guides, and authors of current articles.
3. Send for detailed topographical maps of the areas you wish to visit. The index maps of the United States Geological Survey are free. Write to United States Geological Survey, 1220 Sunrise Drive, Reston, Virginia 22092.
4. Subscribe to the club bulletins of the area you wish to visit.
5. Keep your own index of articles in magazines which pertain to the trips you plan.
6. Look up the names of rock shops in the area you are interested in and order samples of local specimens by mail so that you can study the items you hope to find. A bonus is that if you are a mail order customer of the dealer, the owner may be inclined to give you more help when you get to the shop in person.
7. Besides guide books for individual areas, read books about invertebrate or vertebrate fossils, crystals, agates, geodes, scenic landmarks, and anything else pertaining to goals.
8. If you have adequate safety equipment and a comprehensive liability insurance policy, it can be easier to get permission to collect.
9. It is helpful to plan trips to coincide with local shows, swaps, or field trips.
10. Don't leave things to chance. Plan the important details of your trip in advance before you go.
11. Get lists from the Geological Surveys of the States you plan to visit

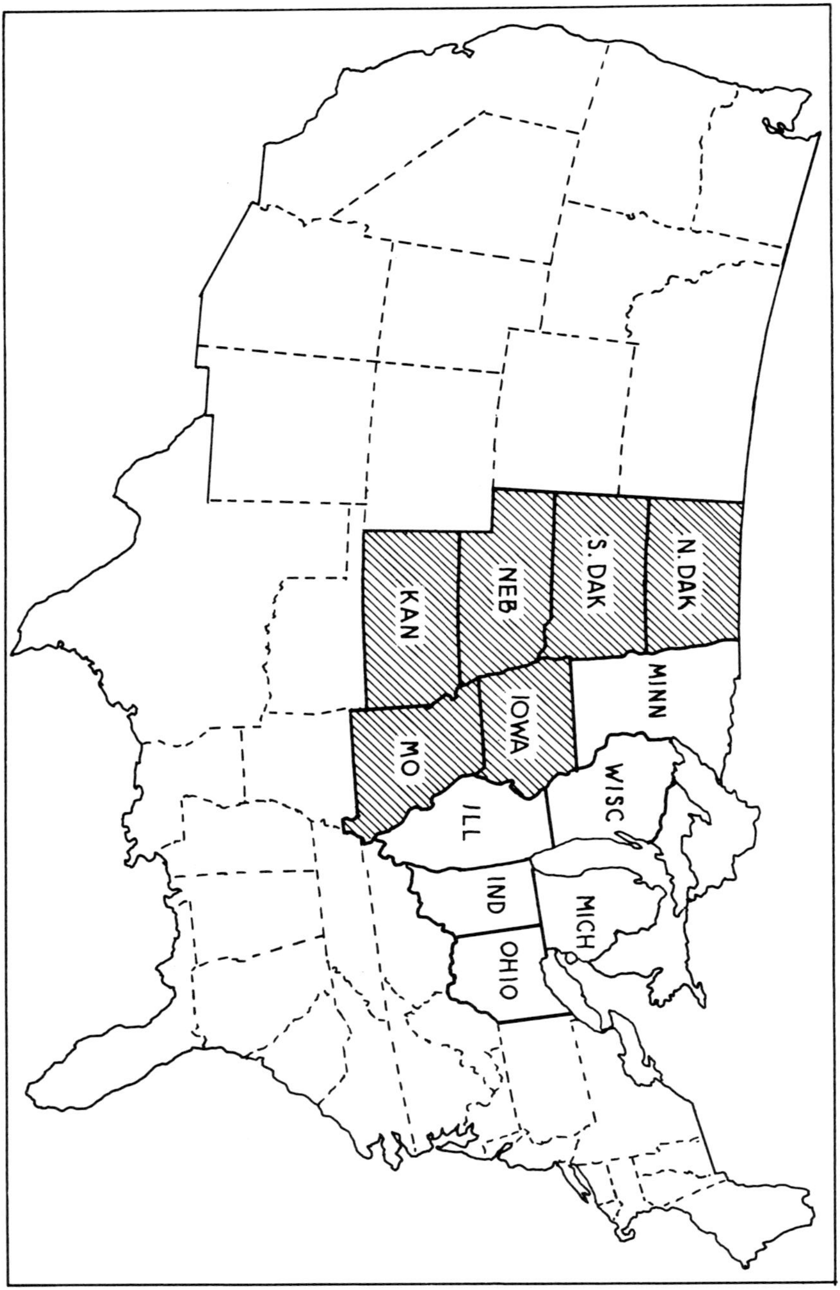

Shaded area indicates 6 states in this book; outlined are the Great Lakes states

of the quarries, mines, and gravel pits in operation and abandoned. The Geological Surveys are located at the Universities in the following cities: Grand Forks, North Dakota; Vermillion, South Dakota; Lincoln, Nebraska; Rolla, Missouri; Lawrence, Kansas; and Iowa, City Iowa. The various states have publications of many kinds, including books about rocks and fossils of the states.

12. The stte departments or divisions of tourism or economic development have free state maps, which are much better than the maps in atlases or from filling stations.

13. If you are members of travel clubs such as AAA or Good Sam, you can get information about places to stay, eat, shop, and sightsee.

14. Member of the club nearest you have probably visited many of the sites you hope to visit. There are also Federation programs about most of the famous localities. Join and support your local club, and you will get lots of help.

In this book the individual state maps have all the counties marked where significant localities occur for gems, minerals, or fossils. Some of these localities overlap into border states. For example the geode area of Iowa and Missouri overlaps into Illinois, and the zinc-lead mining area of Iowa is an extension of that in Wisconsin and Illinois. The states of Illinois, Wisconsin, Michigan, Ohio, Indiana and Minnesota are covered in *Midwest Gem, Mineral and Fossil Trails, GREAT LAKES STATES.*

How To Hunt Minerals, Gems and Fossils in The Prairie States

The six states included in this book are all west of the Mississippi. All of them are at least partially glaciated. All of them have had large areas of prairie, although now most of the original prairie is agricultural land. These states also have miles and miles of waterways, dams, reservoirs, and natural shoreline of the Missouri and Mississippi Rivers and tributaries. Resources of the region are tremendous. All have sand, gravel, coal, limestone, clay and gypsum. North Dakota, Kansas and Missouri have significant deposits of oil and natural gas. South Dakota is the nation's leading gold producer. Other minerals and rocks from the region include quartz, granite, zinc, lead, salt, chert, sandstone, and ores of iron and uranium.

For the rock hobbyist, lapidary, and fossil enthusiast these states have much to offer. Many of the localities have not been hunted as heavily as productive sites on the west or east coasts. Lake Superior agates are widespread. Magnificent fossils occur in each state. Superb crystals of quartz, calcite, pyrite, marcasite, dolomite, celestite, selenite, sphalerite, and barite are not difficult to come by. There are also numerous rare minerals which provide an extra thrill for the field collector.

Climate is an important factor in collecting in any of these states. All of them are subject to sudden extremes and violent storms. Collectors must listen to weather forecasts and heed them. North Dakota has the longest and coldest winters. Iowa has the most snow. Missouri has hot and humid summers. Kansas has the most tornados. South Dakota has frequent droughts. Nebraska has considerable snow, but milder temperatures than the states to the north.

The coolest spot for summer collecting is probably around Custer in the Black Hills. Spring and fall are excellent for collectors in all six states. Winter collecting is possible whenever the temperatures are mild enough and the ground is bare. If you are in doubt about access to the area you wish to visit, call the highway department to determine if the roads are considered safe.

Wherever you travel in this region be prepared for changes in the weather. Sturdy clothing of cotton denim is a good choice. It can be layered with a wool sweater and a windbreaker or parka for cold weather. Comfortable well-built shoes or boots are a must. In some places safety shoes are required, as well as safety glasses and a hard hat. An eye shade or brimmed hat should be worn to protect the eyes from the sun. Gloves are handy when working with sharp rock and heavy tools.

Clothes with plenty of pockets are handy. In the pockets you can store notebook, pencil, tissues, plastic bags, small tools, a canteen of water, a hand lens, and a snack.

If there are several in the group, it is a good idea to carry whistles to communicate when widely separated. Invent a code for messages with whistles. For instance one blast might mean "Where are you ?" and the answering blast means "Over here." Three blasts means "Come at once!"

There are lots of places not listed in this book where you may find excellent specimens, but the book gives you leads for finding these places.

Ask to look over farmers stone piles, amassed from glacial rocks scattered in their fields. Stop and look at any large roadcuts which have not been planted over. Examine new construction sites. Walk slowly along rivers and streams. Stop at gravelled beaches. Don't forget dams and reservoirs.

If there are known deposits anywhere in the area, look for eroded areas or areas of sparse vegetation. Look in the ditches and underbrush. In dry periods look for pot holes and swamps or bogs which may be drying up.

Careful study and determined examination of all possibilities will reward the patient and knowledgeable collector. Learn to be a geological detective and follow all clues with diligence.

It is also necessary to learn to be selective when you do locate a productive area. Take home only what you and your friends can use. Learn to use what you collect to the best advantage.

Many hobbyists have pickups, campers, mobile homes, vans, or trailers. Others use tents for camping in remote areas. Still others drive their regular family cars and stay in motels.

While a passenger car is fine for most field trips close to the beaten path, most cars are not well-suited to long drives on rugged back country trails.

Older model cars have better clearance than late models. Also older models have sturdy frames. Larger tires and overload springs may raise the body of a car sufficiently for rough and rocky unpaved roads. Steel-belted radial tires provide extra safety. Sometimes it is necessary to rig up protection for the gas tank and oil pan in rocky areas. Always carry an emergency kit, water, and a spare tire.

Travelalls, Jeepsters, 4x4s and heavy vehicles often have 2 gas tanks which is handy for remote trips. However in most instances in these states you will not be far from gas stations, garages, and auto stores. In parts of western North Dakota and western South Dakota the dangers of getting lost or stalled require extra planning. Carry extra coolant for your vehicle in the summer, and in the winter have a shovel, a bag of sand and a couple of blankets.

When packing your vehicle, especially after a successful trip, try to store heavy items such as boxes of rocks, over the axles. Heavy boxes should not be put in the trunk. Have maps and guide books in front where they can be easily reached.

All field trips should be guided by basic safety considerations. The land owner or manager may add more rules to be followed. Most accidents are preventable. Be careful and take your time.

Here are some time-tested safety rules.

1. Never work under overhanging rocks or on the edge of a crumbling cliff.
2. Don't go alone on trips to remote and rugged areas.
3. If your vehicle is disabled in a region you are not familiar with and you are far from towns or farms or main roads, it is best to stay with that vehicle. This is especially true in extreme weather.
4. Don't try to work out delicate specimens or find new locations if you are fatigued.

Limestone was quarried for fenceposts in northern Kansas where pioneers found little wood. Many of the stone fence posts still stand. Photo by Kansas Dept. of Development.

5. Don't stand close to a spot where someone else is working.

6. Don't go into old mine workings.

7. Be careful about walking on slippery rocks, or crossing streams or swamps.

8. Never camp beside a flowing stream in a ravine, canyon, or gorge.

9. Do not drink water from a stream no matter how clear it looks. Carry bottled water or water purifying tablets.

10. Light fires only in designated places.

11. Put out all fires entirely.

12. Keep picnic food cold.

13. Stay away from wild animals which could be rabid.

14. Be familiar with the poisonous plants, snakes and insects of each area you visit.

15. Park your car where it is easy to see, and look for landmarks in each direction.

Equipment for a trip depends on the objective of the trip. If the destination is a mine dump to hunt for micromounts, a 10 power hand lens, and some small tools like screw drivers, awls, chisels, and pocket knife should be added to the standard prospector's pick. If the hunt is in a quarry for fossils which must be worked out of thick limestone layers, heavy mauls, crowbars, or mattocks should be included. For stream hunting, a gold pan and gem screens will add to the excitement. A long-handled tool is convenient for turning over rocks in gravel beds to see if a delightful pattern is lying face down.

Optional equipment also varies with the purpose of the trip. A hardness kit helps with field identification. A pump or spray can or bottle of water may help decide whether a piece of jasper or flint will be improved by polish. Epoxy, white glue, or shellac, soft brushes, bent wires, and burlap may be needed for fossils. Proper wrapping supplies are always needed.

Other sensible supplies are matches, water, insect spray, whistles for signals, and a compass.

The long field trip should be planned with care. The most essential preparation, getting current information on the sites to be visited, is often neglected. Get this from clubs, dealers, guides, Chambers of Commerce, mining companies, fee areas, and Tourist Departments. Accurate maps are required. Write the U.S. Department of Agriculture, Washington, DC 20250 for National Forest Service maps. For BLM maps and information write to the U.S. Bureau of Land Management, 7981 Eastern Avenue, Silver Springs, MD 20910.

Above all, any collector who visits any of these states should always remember that future generations should also have access to these sites, and that whether or not these sites stay open depends on how he conducts himself. It has been a long hard struggle with government agencies, with mining companies, and with land owners, to have as many collecting spots open as we still have. Many hobbyists have worked hard so that today's hobbyists may have opportunities to collect. Collecting is a privilege, not a right. Make every trip a model of outdoor sportsmanship, ethics and manners. Add to our reputation as serious amateur scientists, not vandals.

Public meetings and comment periods are often held about land status. Attend these or write letters. Cooperate with other states where withdrawals of sites are threatened.

The American Federation Code of Ethics for collectors is a good guide to follow. So too is the Golden Rule.

THE AMERICAN FEDERATION CODE OF ETHICS

I will respect both private and public property and will do no collecting on privately owned land without the owner's permission.

I will keep informed on all laws, regulations or rules governing collecting on public lands and will observe them.

I will, to the best of my ability, ascertain the boundary lines of property on which I plan to collect.

I will use no firearms or blasting material in collecting areas.

I will cause no willful damage to property of any kind - fences, signs, buildings, etc.

I will leave all gates as found.

I will build fires in designated or safe places only and will be certain they are completely extinguished before leaving the area.

I will discard no burning material - matches, cigarettes, etc.

I will fill all excavation holes which may be dangerous to livestock.

I will not contaminate wells, creeks, or other water supply.

I will cause no willful damage to collecting material and will take home only what I can reasonably use.

I will support the rockhound project H.E.L.P. (Help Eliminate Litter Please) and will leave all collecting areas devoid of litter, regardless of how found.

I will cooperate with field trip leaders and those in designated authority in all collecting areas.

I will report to my club or federation officers, Bureau of Land Management, or other proper authorities, any deposit of petrified wood or other material on public lands which should be protected for the enjoyment of future generations for public educational and scientific purposes.

I will appreciate and protect our heritage of natural resources.

I will observe the "Golden Rule," will use "Good Outdoor Manners" and will at all times conduct myself in a manner which will add to the stature and "Public Image" of rockhounds everywhere.

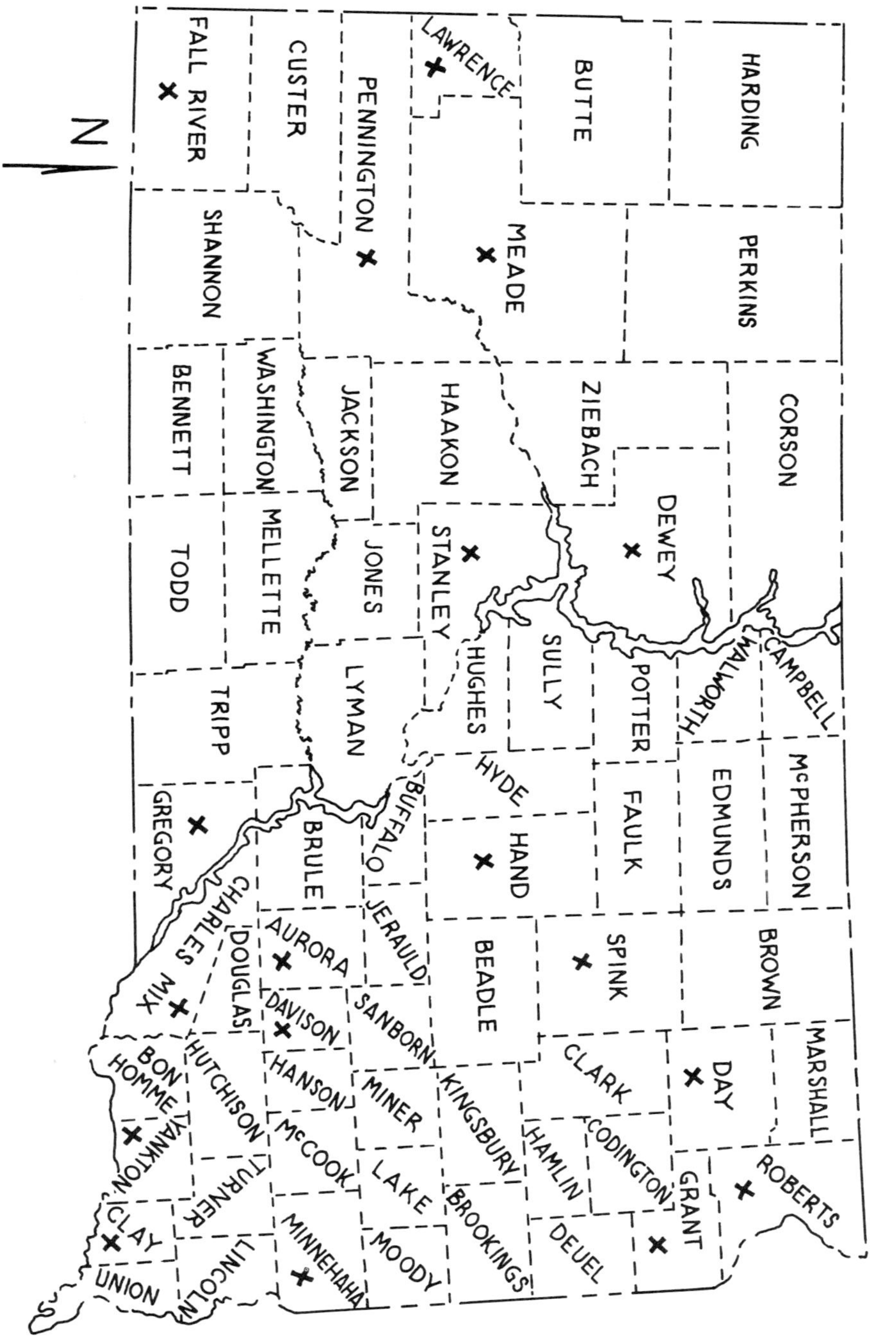

Xs indicate counties with locations for fossils, gems and minerals

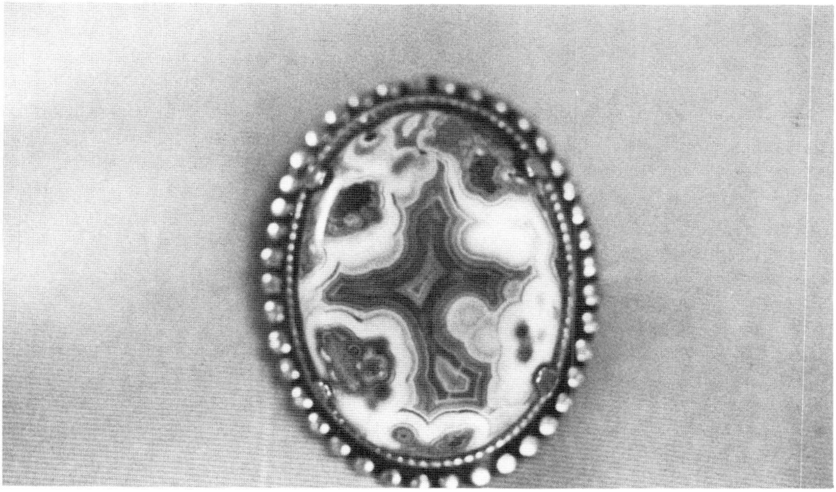

Tepee Canyon agate from the Black Hills is popular for original jewelry.

SOUTH DAKOTA

South Dakota is the extreme western boundary of a large region known as the Midwest. It is a state of varied landscapes, with the rugged Black Hills rising over 7,000 feet in the southwest, the arid sculptured badlands and the buttes of the Missouri in the center of the state, and the flat plains and rolling prairies covering the east. The state has the special distinction of having the geographical center of the United States just seventeen miles west of Castle Rock.

South Dakota should be visited by the rock collector in the spring or fall. Late May and June are nice months, surpassed only by September and October. July and August are very hot in most collecting sites. If it is necessary to come during those months, the early morning hours are the most pleasant. Summer days are long, so collecting from about 5:00 PM to 8:00 PM is also productive.

Western South Dakota

Thousands of acres of collecting possibilities can be found in the Black Hills, the badlands, and the National Grasslands. Many good locations are on Public land.

Widely scattered alluvial deposits of the National Grasslands have arid and rough terrain. Some of the land is fenced in for grazing. The Buffalo Gap National Grasslands headquarters is in Wall, South Dakota, where you may purchase a map showing where rock beds are located. No permission is

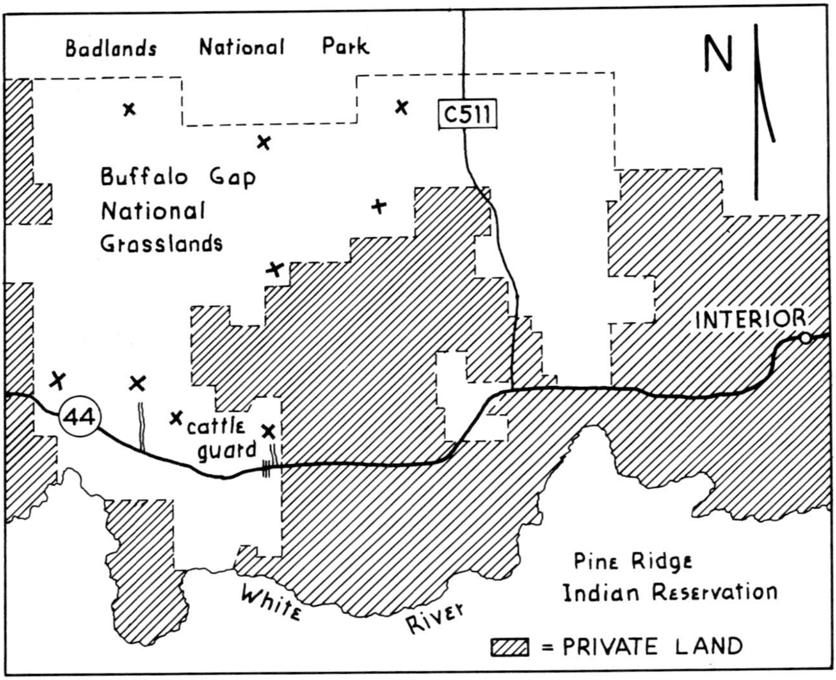

Xs show Conata agate beds in Buffalo Gap National Grasslands

necessary to collect on these lands. The office at Wall has samples of the cutting materials and fossils of the Grasslands.

Best-known of the rock beds of the National Grasslands is the Conata area. A rough area just south of Badlands National Park, the Conata rock hunting site is west of the town of Interior, located on SD Highway 44. At 7 miles, County Road C 511 will be on the right. One-half mile father is a sign Conata East. The first beds are north of this spot. Four miles farther on Highway 44 is a cattle gate on the north side of the road with a trail leading north to many scattered rock beds. There are large bare badlands buttes close to the main road here, but the trail leads through grassy areas to the rock beds. Look both east and west of the trail.

Don't try to test your car by driving over badlands formations, and do leave the area in a hurry if it starts to rain. The gumbo soil is as sticky as glue and dries like concrete.

The Weta rock beds are farther to the east. From Highway I-90 take the exit for 73 north, a State Route going to Philip. The first road south of this exit and parallel to I-90 is an east-west route, old U.S. 16. Turn west and go 1 mile. Turn south here on a narrow country trail which takes you through some badlands formations and across a railroad. Cross the railroad and continue south for about 2 miles where the main trail turns west for a short distance. At the next fork turn south again. In another 2 miles the trail turns west for a mile, and then south once more. You will soon see large rock beds on both sides of the trail. The beds continue all the way to the White River. The grasslands are not fenced, but there is some private property near the river which is fenced. The grasslands areas furnish many thousands of acres of good collecting and no permission is required.

There are pieces of jasper, petrified wood, silicified marine fossils, puddingstone and breccia. The puddingstone, technically jasper conglomerate, is a good lapidary material with multi-colored rounded pebbles embedded in a hard matrix. The breccia inclusions are angular. There are also many colors of quartz, including rose, yellow, white and smoky.

Chalcedony in the grasslands is in veins. Sharply broken sections are easily distinguished from water-worn agates. The chalcedony is found over a wide area in many colors. Sky blue is the rarest. There is also a "sandwich" chalcedony, blue with pink, white with red, and black with white. Much of the chalcedony is fluorescent.

The rarest of the agates is the Fairburn. More about this later. Fairburns are difficult to spot among the other rocks of similar shades, colors, and patterns, but many have been found in remote areas as in the original beds.

Collectors should especially look for jasper and agate with imprints of gastropods, brachiopods and other fossils. Fossil "sponges" and horn corals are local favorites. Specimens of wood should be closely examined, because rare and exotic cycad is found in these badlands areas. Other rare fossils more common here than elsewhere are fossil eggs. The birds eggs have been replaced by chalcedony.

Black agates are found 1.5 miles west of Scenic. East of Scenic near the

Xs indicate Weta agate beds of Buffalo Gap Grasslands

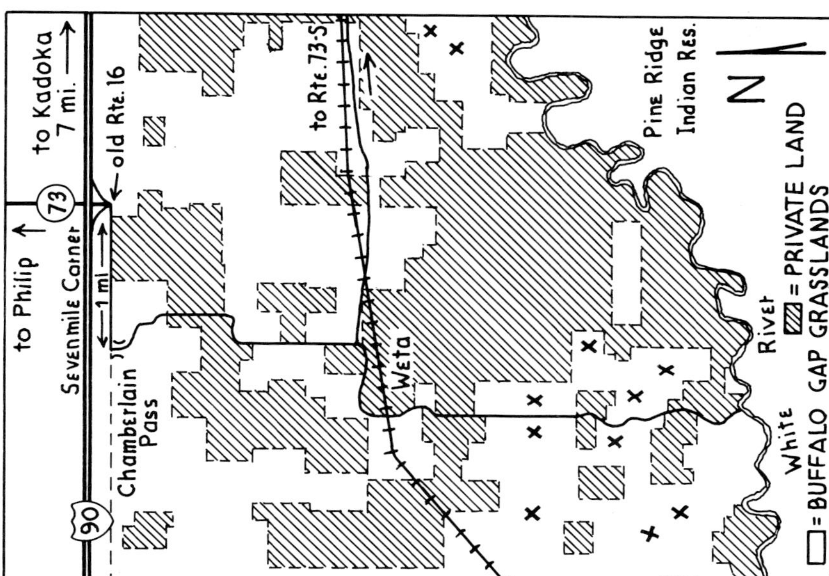

Two rare finds in the badlands are agatized bone and agatized birds egg. Zeitner Collection, Pioneer Museum, Murdo.

railroad track at the former site of Imlay, small chalcedony geodes are found, lined with sparkling drusy quartz crystals. A few are spotted with pink or orange.

Along with all these quartz gems are some other collectables. There are selenite crystals, nut-shaped or button-shaped barite nodules, and pieces of pink or red chatoyant feldspar.

One of the brightest and most boldly patterned of all American agates was discovered near the small town of Fairburn in Custer County, South Dakota. To this day, although the agates are found in numerous other spots in South Dakota, Nebraska and Wyoming, they still bear the name of Fairburn.

The official State Gem, these vivid fortification agates are brightly banded in several colors. They vary in size from about an inch to 10 inches in diameter, but most of them are only 1.5" to 2.5" across. A good quality Fairburn the size of a hens egg is very rare. Nevertheless fine specimens are found every year by patient, devoted, and lucky collectors.

Since several other materials in the vicinity are brightly colored and banded, the novice has great difficulty recognizing Fairburns in the rough.

The fortifications (bands) of the Fairburn are extremely sharp and complex, resembling a thorny holly leaf. These bands are in contrasting colors and are exactly parallel. If the lines are not sharp and the bands are not parallel, but "hit or miss", the agate is not a Fairburn. The failing candidates include the Prairie agate, which has bright swirly irregular lines, and the "big deal", a piece of jasper with a small area of fortifications limited to one or two spots. A real Fairburn is all agate, although it can have a quartz center. It is translucent when cut, and the pattern goes all the way through. Sometimes there is little hint on the outside of the glories of the agate on the inside. These are pockmarked and usually dark. The full face pattern is easier to recognize. It is so distinctive it needs no expert to identify it.

According to legend, an agate fancier from a wealthy Michigan family came to the area many years ago. He marked off the Fairburn rock beds into small squares with stakes and ropes, and then hired all the children in the region to take a square at a time and turn over every individual stone. The man was supposed to have taken away bushels of choice agates. This highly unlikely myth vies with one about a basement under the ruins of an old house, full of wooden boxes of Fairburns.

The first authentic story is about the "Kern" beds, the original location. "Grandma" Kern lived alone, and loved the beautiful bright rocks she found on her property. She spent most of her time looking for the agates, and kept neighbors and visitors away with a shotgun. As she grew older, "Grandma" began losing her sight, so she became somewhat more cooperative. She allowed rock hunters on her property, and followed them around wearing a nail apron full of silver coins, which she eagerly exchanged for the agates the collectors found.

Many ranchers in the Fairburn and Hermosa area have excellent collections. Most of the old collections have been sold, some in their entirety and some piece-by-piece to the highest bidders. To people from other states the prices of good Fairburns are "scandalous". I have seen agates no larger than an egg sell for $300.00. South Dakotans know how hard it is to find these prizes, and they dearly love these agates, so really fine examples are seldom for sale.

A novice at this type of hunting should study the South Dakota agates and gemstone materials at the museum of Geology at the South Dakota School of Mines and Technology in Rapid City.

The town of Fairburn is just off S.D. Highway 79, south of Rapid City 25

Old Fairburn beds

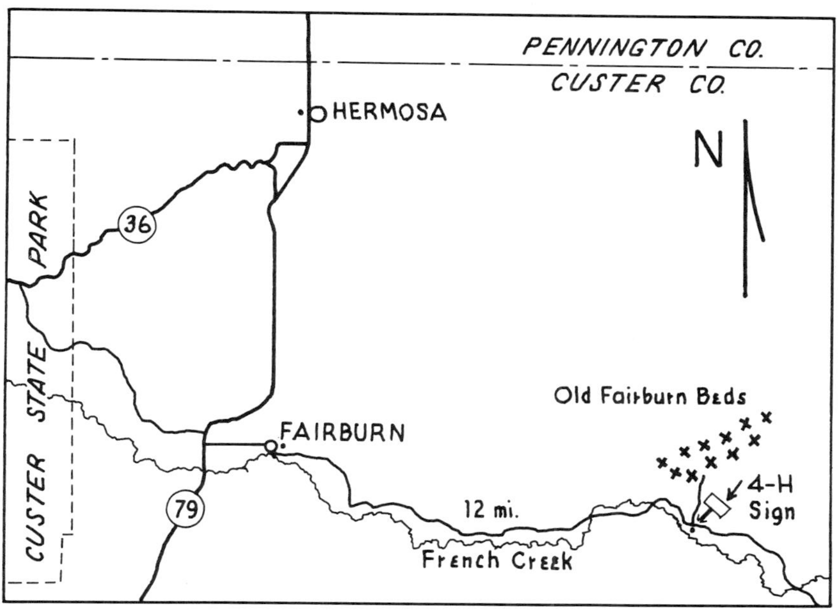

Fairburn agates are among the most colorful fortification agates. Photo by Don Goodart, courtesy of South Dakota Museum of Geology.

miles. A sign points to the town. The principal road going east of Fairburn is French Creek Road, a dirt and gravel road maintained by the county.

Take this winding and hilly road east following the main tracks 12 miles east. Turn off here on the trail to your left where the 4H club has maintained a picnic area. Park in this area and walk into the intriguing buttes, ledges, and washes, where thousands of choice specimens have been found.

There are rattlesnakes in South Dakota. Although we rarely see one on a rock trip, it is possible to run into a nest of them anywhere in the west. Wear boots, make a noise when you walk, and don't reach into any spot you cannot fully see.

There are other agate beds with similar quartz family gems scattered all the way south into Nebraska. Some of these are near Red Shirt, Indian Creek, Railroad Buttes, Creston, Kadoka, Ardmore, Smithwick and Oelrichs. When surveyed by plane, all rock beds had vehicle tracks leading to them, but in spite of the fact there are no secret places, good agates are found every year.

Guides are sometimes available to take collectors to remote areas by Jeep. Inquire at the Chambers of Commerce in Wall, Hill City, and Hot Springs, at the Skyline Rock Shop in Rapid City, or at the Rock shop in Okaton. Two Guides who have led trips for many years are Ken Renner, retired ranger, Box 95, Wall, SD 57790 and Louis Stratton, Box 695, Custer, SD 57730.

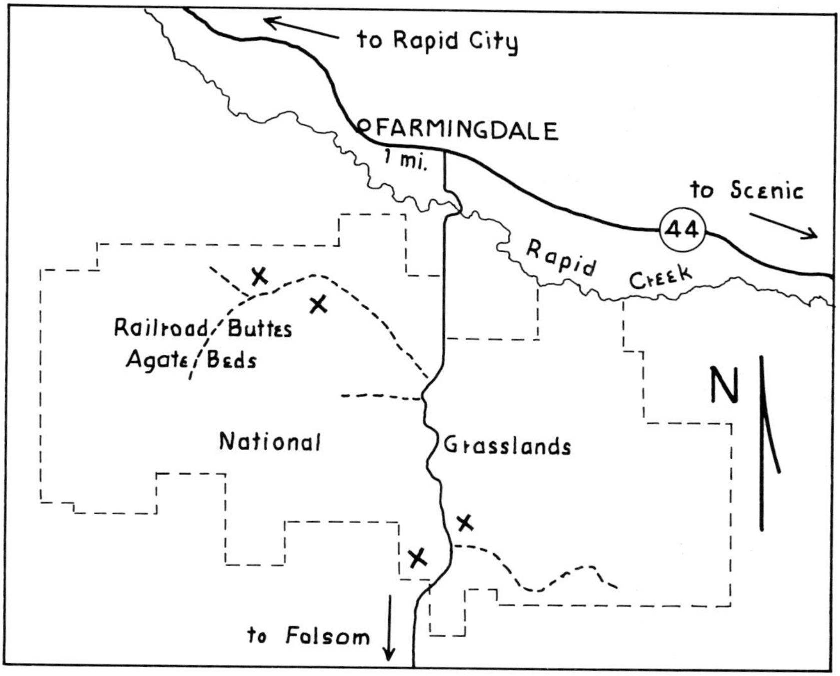

The Kadoka agate beds in National Grasslands

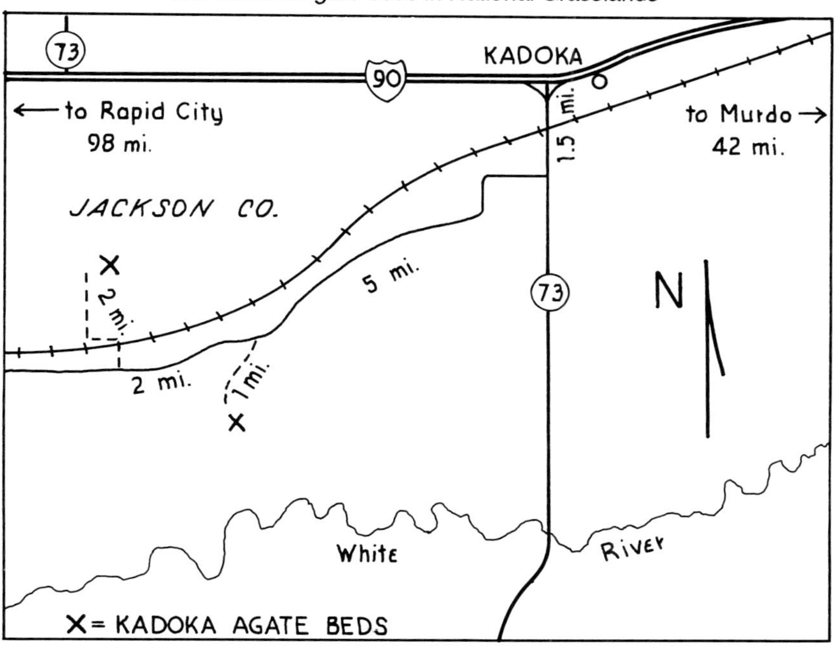

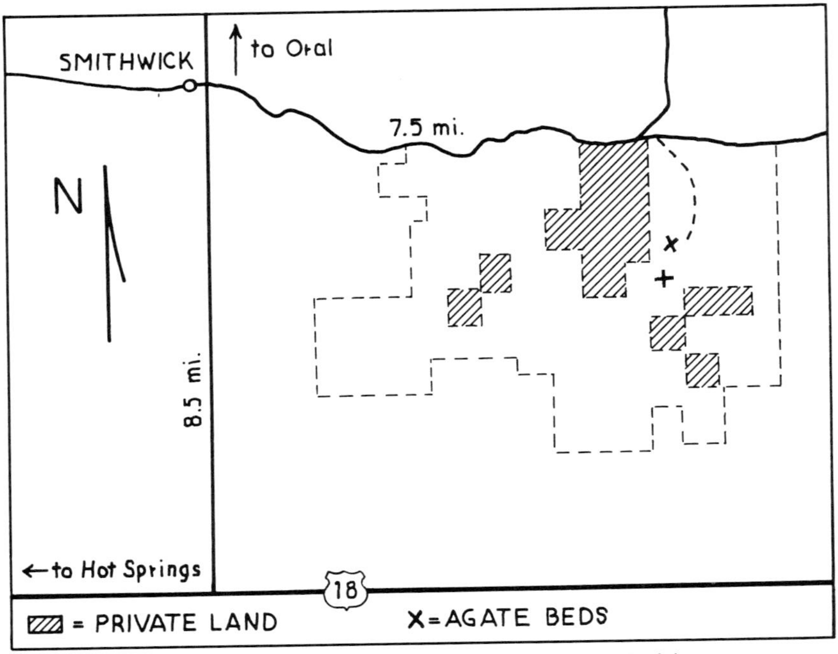

Agate beds in National Grasslands near Smithwick

Agate beds near Oelrichs on National Grasslands

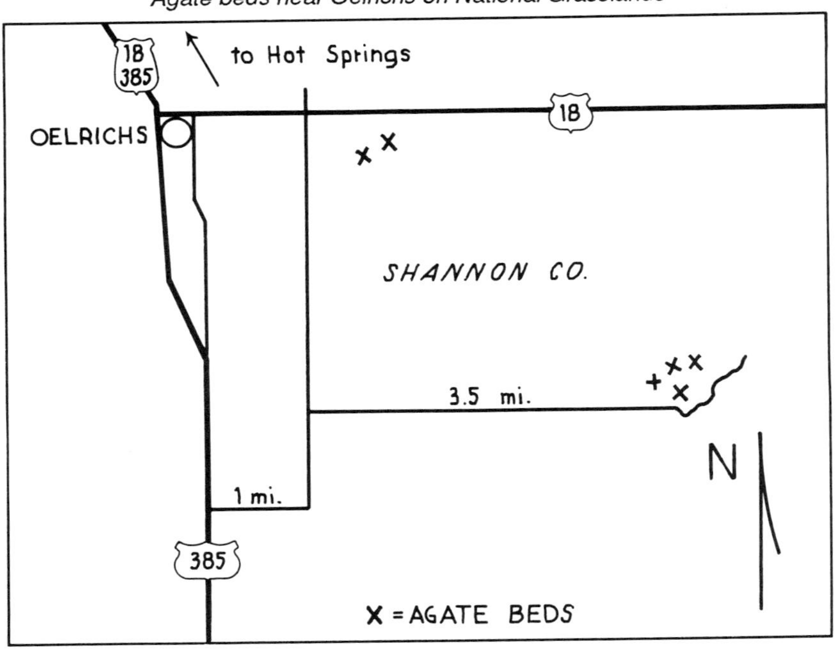

Well-formed transparent golden barite crystals have been found in several places in western South Dakota and are considered among the finest barite crystals in the world. According to John Sinkankas they are even facetable. Occurring in large concretions in the Pierre Shale, the crystals are associated with showy lemon-yellow calcite crystals, and velvety siderite. The classic locality, along Elk Creek, Meade County, is pretty well worked out. However, the big concretions continue to become exposed and the area where they occur is about 20 miles long from the confluence of Elk Creek and the Cheyenne River to Viewfield north of New Underwood. This is all private property so permission must be obtained.

Another barite locality is in Hay Draw 10 miles southwest of Union Center and other draws of the region. Similar to the Elk Creek barite, these crystals are also in concretions.

Esthetic barite crystals and crystal groups occur in Pennington County on Boxelder Creek north and northeast of Owanka. Sometimes the cephalopods which occur in the same concretions are lined with crystals of barite and calcite. Other barite has been reported west and south of Scenic.

Colorless tubular lustrous crystals of barite occur in Jackson County, south of Belvidere and north of the White River. The location is south and west of old U.S. 16.

In Brule County barite rosettes occur. They are gray in color and up to 6

Xs indicate barite locations in South Dakota

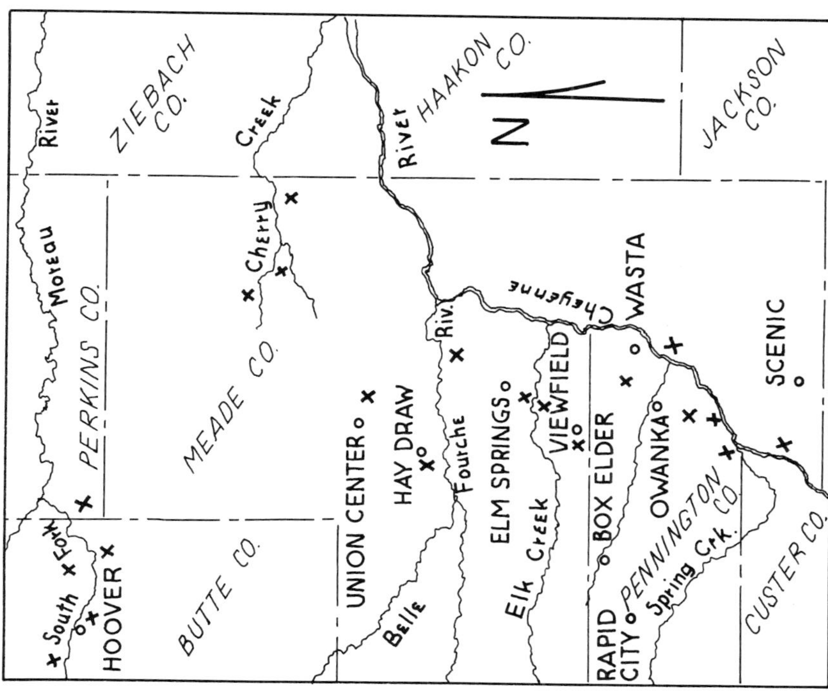

Rock beds in eroded badlands terrain stretch for many miles in Southwestern South Dakota. Photo, South Dakota Department of Tourism.

Ancient sea life, probably a coral, has turned to bright agate. Zeitner Collection.

inches in diameter. The most numerous finds are on Elm Creek near its junction with the Missouri River 15 miles south of Chamberlain in Pierre Shales exposures.

Elegant curves of rams horn gypsum are found with the barite in concretions near Hoover in Butte County.

THE BLACK HILLS

In 1876 the Black Hills was the scene of a frenzied and flamboyant gold rush. Placer miners rushed to the new field from all over the west. They worked every creek, then gold-bearing lodes led to underground mining. The most famous of these old mines, The Homestake in Lawrence County, is still in operation and is the largest producing gold mine in North America.

In recent years there has been renewed interest in gold mining in the Black Hills. Many new claims have been filed. Every summer grains and flakes and nuggets of gold are found by amateurs panning the gravels of scenic streams, flowing through the mountainous Ponderosa Forest.

Panning is allowed in the streams of the National Forest. Some of the best areas are in Pennington County and Custer County. Streams near the communities of Keystone, Silver City, Rochford, Rockerville, Hill City and Custer attract many part-time prospectors.

Gold in the stream is yellow, but without a brassy luster. It is soft and malleable. Look-alikes in the sand, which may confuse the beginner, are yellow mica and pyrite. The mica is too hard and brittle to be confusing for very long, and it is stacked in perfect horizontal cleavage planes. The pyrite is hard and looks like angular pieces of tarnished brass.

Several places offer fee basis panning in salted gravel. information is available from Chambers of Commerce. Gold pans and equipment are available in area rock shops.

Forest Service maps may be obtained from the Ranger Stations in Rapid City or Custer. Look for sections of stream which have a convenient trail nearby where you can park and work on Forest Service land. If the maps show mining symbols near the streams, that is a good sign. Gold has been found in all the major streams of the Black Hills. Some streams are almost dry during late summer. Best-known gold-bearing creeks are Spring Creek, Castle Creek, French Creek, Iron Creek, Battle Creek, Boxelder Creek, Elk Creek, Grizzly Creek and Squaw Creek.

The pegmatites of the Black Hills constitute one of the prime areas in the United States for a wide variety of minerals. Over 20 new minerals have been discovered in old mining districts here in recent years. Larger and better crystals of previously known Black Hills minerals have also been found.

"Black Hills" is really a major misnomer. These are mountains, not hills, and they are green, not black. This is an ideal area for a summer vacation for an amateur mineralogist. The park-like Ponderosa Forest is liberally sprinkled with old mine dumps.

Pegmatite is a coarse granite, occurring in the form of small or extensive dikes. Having cooled slowly, the component minerals of ordinary granite have crystallized separately. The pegmatite dikes of the Black Hills have been mined for feldspar, quartz, beryl, mica, spodumene, and other minerals. Many of the minerals found here are extremely attractive to collectors. There are red garnets, blue apatite crystals, lavender lepidolite, blue and green tourmaline, and dark columbite and tantalite. There are blocky staurolites, lacy aragonite, sharp calcites, heavy cassiterites, and rare phosphates. There are over 300 known minerals found in the Black Hills.

Micromounters probably have a better chance of coming up with the answer to their dreams of discovering a new mineral in the Black Hills than anywhere else in the United States. Willard (Bill) Roberts, a gifted mineralogist who started as a rock hobbyist when he was a boy, was responsible for finding, identifying and describing many of the new minerals. Willard had help from his associates and students at South Dakota Tech in Rapid City, from Dr. Paul Moore of Illinois and other eminent mineralogists. Among the 22 new species were some complex phosphates. Bill also found 100 minerals new to the Black Hills.

Among the new species which have been accredited and described are robertsite, olmsteadite, and tinsleyite, all named for their finders, sinkankasite, named for author John Sinkankas, tiptopite named for the mine near Custer, and jahnsite, selegerite, cernyite, perloffite, ehrleite, walentaite, whitmoreite, fransoletite, griphite, kehoeite, maghagendorfite, rosemaryite, metauranocircite, and ferro-wyllieite. Roberts was working on other possible new species when he died.

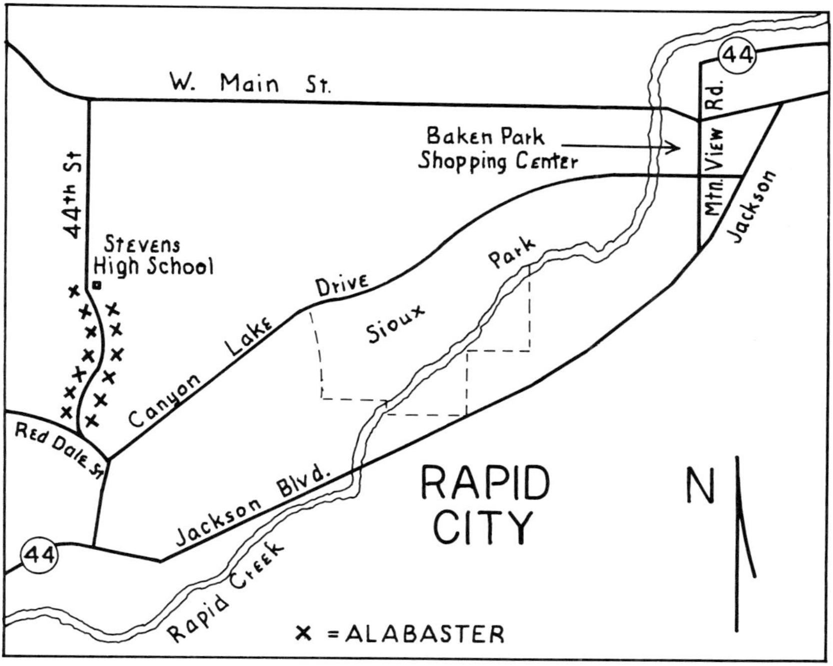

Alabaster locations in Rapid City

Roberts was active in hobby as well as professional associations, and had a great influence on the gem and mineral collectors of the nation, particularly the micromounters. Roberts was also co-author of the immensely popular *Encyclopedia of Minerals.*

Only a few of the old mines are still operating. These are near Custer, Keystone, and Lead-Deadwood. Old prospects which have reverted to the Forest Service may be hunted without permission. Still, some of the old mines are private property and are posted. If the dump you wish to hunt is fenced, posted, or marked with valid claim markers, you will have to get permission from the owner or caretaker.

The mines have furnished commercial minerals as well as fine specimens for museums, private collections, and micromounters.

Some of the famous locations are in Pennington County.where there are large gravel deposits. White and mottled alabaster is abundant in and around Rapid City. Road cuts near Stevens High in West Rapid City expose piles of it.

The Bob Ingersoll Mine, 2 miles west of Keystone, is nationally recognized for an exceptional variety of minerals. Besides gemmy green and blue tourmaline, it has produced blue and pink beryl, lavender lepidolite, white microcline, yellow-green autunite and shiny cleavelandite, to name a few. The mine dumps are occasionally open to students, teachers, scientists and groups.

The Etta Mine, Keystone, is world famous for its gigantic spodumene "logs" and also for an amazing variety of associate minerals. Ludllamite, torbernite, siderite, rutile, kaolinite, and chalcopyrite are some of the minerals found here.

South and east of Keystone is the Hugo. Look here for pink beryl, pink and white muscovite, amblygonite, fluorapatite, and black tourmaline (schorl) and more. The Hugo has been the largest microcline producer in the region. On Iron Mountain Road south of Keystone (Alternate U.S. 16), a rose quartz outcrop is on the left near the top. Limonite, some of it iridescent, is plentiful. It has been mined for specimens.

The Dan Patch Mine, 1 mile west of Keystone, has sphalerite, muscovite, triphylite, columbite-tantalite, albite and beryl. Some of the muscovite is in attractive crystal groups.

The Wood Tin Mine is 1 mile southeast of Keystone. This old mine has produced cassiterite, autunite, white beryl and albite, red and green tourmaline and pink microcline.

Just southeast of Keystone on a gravelled road, is the Madison Mine, which is celebrated for columbite-tantalite, cleavelandite, blue apatite and blue beryl, black tourmaline and rose quartz.

The Peerless has quartz crystals, rutile, andalusite, peach-colored microcline and lavender apatite. This mine is at Keystone. Two other fine Keystone mines are the King Lithia and the White Cap. The White Cap minerals include ruby sphalerite, gemmy violet apatite, and azurite and malachite.

Malachite and azurite as well as other copper minerals, plus native copper, have been found at the Blue Lead Mine on Blue Lead Mountain south of Sheridan Lake.

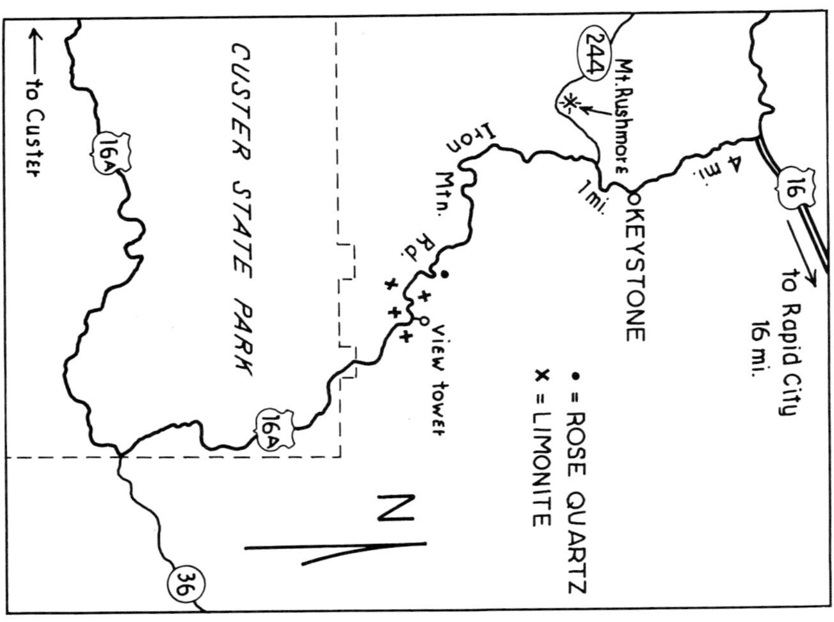

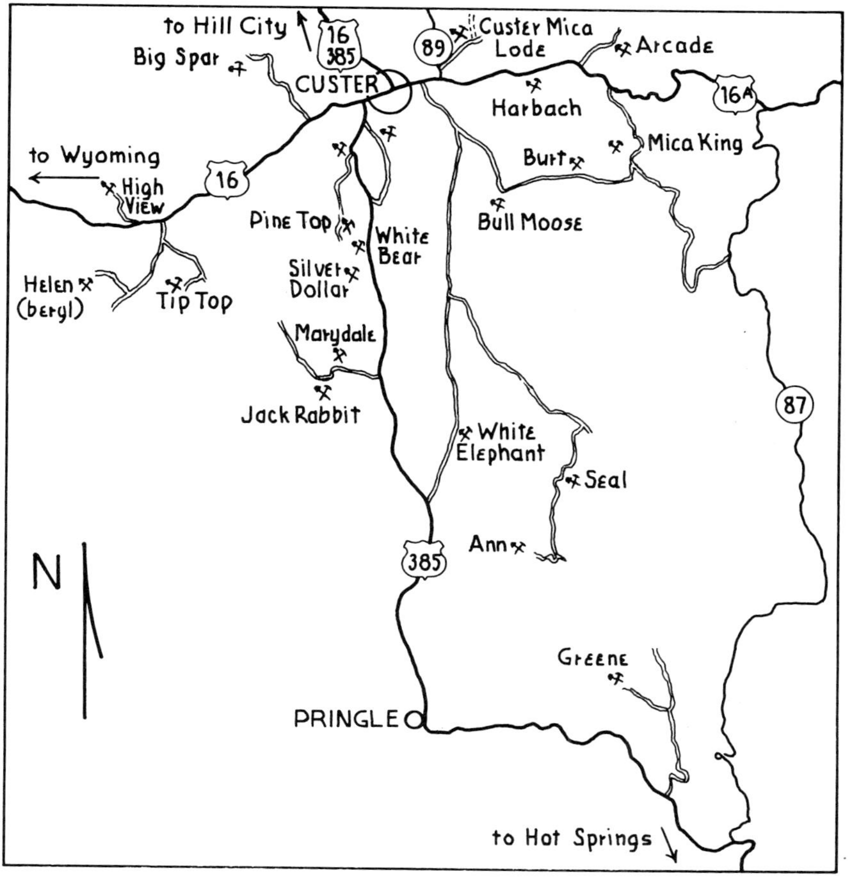

Some of the mines near Custer for quartz, mica, beryl, apatite, tourmaline, feldspar and other minerals

A fabled mine in Custer County is the Helen Beryl Mine, 6 miles southwest of Custer. The beryl comes in several colors and has been found in large crystals. Spodumene from this mine has been in the form of the green gem variety, hiddenite. Some of the pink microcline is chatoyant.

Tin Mountain Mine, 7 miles west of Custer, off Highway U.S. 16, is an excellent mine for a great variety of minerals, including microcline, malachite, azurite, lepidolite, cassiterite, autunite, beryl, and fluorapatite.

An 18 ton mass of columbite-tantalite was found in the Beecher Lode 5 miles southeast of Custer. Sillimanite, green apatite and andalusite have also been found here.

The listing of these Black Hills Mines and their minerals is to show the tremendous variety, and to give approximate locations where prospectors might look for specimens. The mention of these mines does not give current status, as that status changes rapidly. Nor does the mention of any mine indicate that collecting is permissible now. Use maps, and inquire locally.

Take the usual precautions while working in the dumps. Since most of these mines are very old, do not go near any underground workings. On most dumps you will find muscovite; white, brown, or black mica; white, smoky, or rose quartz; black tourmaline; salmon, pink, red, white, or greenish feldspar; limonite, hematite, galena, pyrite, garnet, and more. Although most crystals are very small, a 42 foot spodumene crystal and a 40 ton beryl crystal are among the records.

For your own exploration take along a good hand lens, a prospector's pick, some pry bars and hand tools, and a fluorescent light. The keys to mine dump hunting are a knowledge of the associations of minerals, and persistence. You will find quartz, feldspar and mica and if you are lucky you might find aquamarine, golden beryl, watermelon tourmaline, or even a new mineral. Following are two specific sites for collecting in the Black Hills.

Gemmy, small almandite garnets are found near Tin Mountain, 7 miles west of Custer. Follow U.S. 16 for 7 miles and turn right on a Forest Service trail. A Forest Service sign there may read "Deer Camp", "North Pole Camp", or something similar. Follow the main tracks north until a steep rocky ridge is close to the trail on the right. Park here and climb the ridge. Thousands of glittering garnets are embedded in the schist. The larger garnets will be found in the contact zone of the schist and quartz. Small garnets washed out of their matrix are also found in the dry wash on the left side of this trail.

One of the brightest and most distinctive agates of the west is Tepee Canyon agate. Rivaling the Fairburn in color and pattern, the Tepee agates are found in-place, in the limestone walls of a dramatic canyon. There are several claims in the Canyon which may be valid, and they are marked. It is best to

Mineral and gem locations near Custer

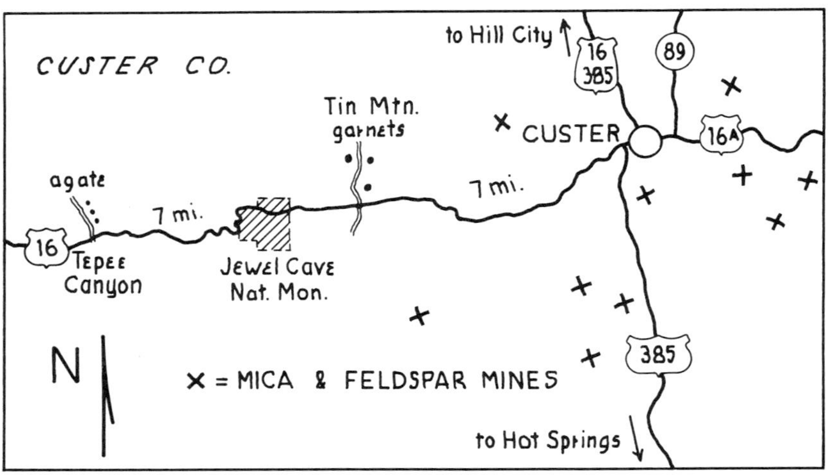

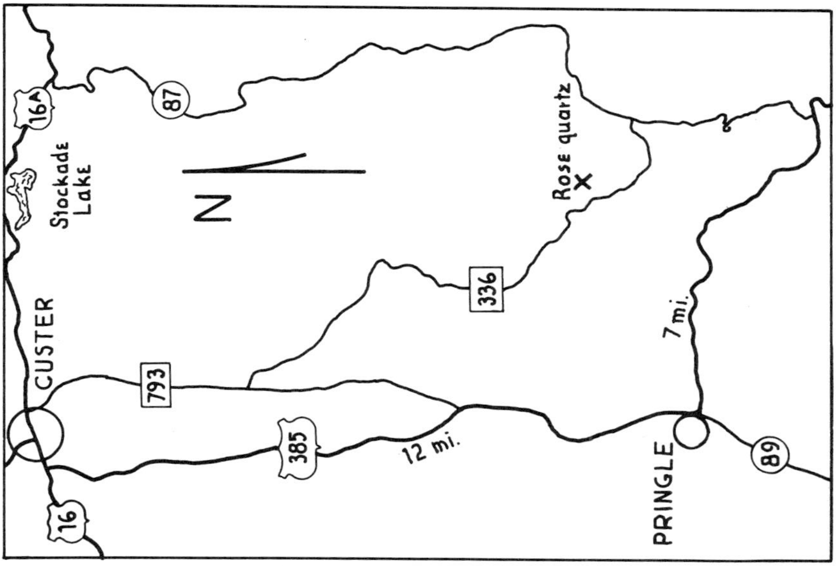

Rose quartz east of Pella

check in Custer at one of the rock shops before you go. Mr. Louis Stratton may still offer guided trips. One of the claims may be open on a fee basis. Most of Tepee Canyon however is Forest Service land, and the agates are not confined to the claims which are close to the road.

Tepee Canyon is 14 miles west of Custer on U.S. 16, a mile beyond Jewel Cave National Park, one of the most amazing of all caves. The Tepee diggings will be seen on your right. There is a Forest Service trail here. Turn right on this trail and look for a place to prospect.

The large nodules of jasper and agate have fortification patterns of red, yellow, orange, and purple. The limestone matrix will polish, but the agates are more attractive if cut out of the matrix.

Be especially careful not to work under an overhanging ledge of rock. These overhangs are lethal. Besides watching above for falling rocks, watch out below for rattlesnakes.

There are more deposits of rose quartz in South Dakota than anywhere else in North America. The area around Custer is particularly known for this lovely variety of quartz. Recently portions of large rose quartz crystals have been discovered in a mine near Custer. Rose quartz crystals have previously been found only in Maine and Brazil. The Scott rose quartz mine in Custer County is a major world producer of this decorative and ornamental material. This is an active mine, privately owned.

Another location is near Pringle, south of Custer on Forest Service Road 336, just off U.S. 385.

Rose quartz also occurs in the following mines: Ross, White Elephant, November, Wylie, Apex, Arcade 1 and 2, Pine Top, Big Chief, Bull Moose, Bull Elk, Tip Top and Sutherland.

There are many locations for various types of barite crystals in western South Dakota Museum of Geology

The quartz varies in color from pale pink to rose pink, deep rose, and lilac-pink. It is sometimes transparent, sometimes translucent and sometimes opaque. An unusual occurrence is silky rose quartz which will cut star stones and star spheres. One location for the asterated rose quartz is the Wiley Mine 1/4 mile east of Highway 385 a few miles south of Custer.

The color of rose quartz is due to titanium and manganese. The asterism in some specimens is caused by oriented fine needles of rutile.

There is a possibility of finding rose quartz in almost any pegmatite in the Black Hills. Dendritic rose quartz, or rose quartz with black tourmaline crystal inclusions make outstanding specimens.

Southeast of Custer ruby red muscovite is the unusual product of the Glenwood Mine. Facet quality aquamarine has come from the Elkhorn Mine, also southeast of Custer.

A third major mining area of the Black Hills is Lawrence County in the northern part of the uplift. The mineralization is quite different here than in the southern hills, and many of the old mines are now owned by Homestake Mining Company, with headquarters at Lead. There is sill some collecting in road cuts, railroad rights of way, and on Forest Service land. Also collecting on privately owned land is possible with permission.

The Homestake Open Cut is a fascinating site to collectors. A tour of the Homestake is available in the summer months.

Lawrence County has geodes in two locations, Spearfish Canyon, near Bridalveil Falls, and on Whitewood Creek. In these geodes are inclusions of fluorite, calcite, hematite, siderite, ankerite and occasionally amethyst. Most of the geodes are quartz.

Strawberry Hill, 5 miles southeast of Lead, is a location for hematite. Bald Mountain has attractive dolomite. Cerussite and galena come from the Iron Hill Mine of Carbonate. The Clover Leaf Mine at Roubaix has galena, labradorite and hornblende. The Hidden Fortune Mine, Lead, has blue barite, currusite, calcite and wulfenite. Some of the crystals are "micro" in size. Celestite, scarce in the Black Hills, comes from the Hidden Treasure Mine.

Hints for Collecting Minerals

1. Necessary items are prospector's pick, safety goggles, gloves, wrapping materials, hand lens, notebook, pen, hand weeder or hand rake. Your equipment should also include items from the following categories and possibly some of the options.

   Shovels: collapsible shovel, spade, miner's shovel

   Large picks: miner's pick, 16" ore pick, mattock

   Striking tools: crack hammer, chisel point pick, maul, long striking hammer

   Chisels: cold chisel, long chisel, wedge, gad, screw driver

   Options: hydraulic jack, screw jack, augers, hand drill, gold pan, gem screens, ultra-violet light, bent wire, paint brush
2. When working in a mine dump look for any mineral which is different from the rest of the pile in color, translucency, shape, luster, or texture.
3. Look for specimens which are combinations of several minerals.
4. Work with someone. When there is heavy digging or rock moving, alternate jobs.
5. Look for cavities in the rock walls.
6. Split large rocks which are composed of several minerals.
7. Look for a contact zone, an area where two different types of rock meet.
8. Micromounts are found in small seams, vugs, old natural fractures, between mica and feldspar plates, and in loose coarse material. Examine all specimens with suspected micromount qualities with a 10 power lens.
9. If you find a good specimen, try to trace where it came from.
10. In any mine or quarry identify the principal rocks, know what minerals may be found with them, and seek out any layer which shows the characteristics you are looking for. Prospect several areas quickly before selecting a place to dig.
11. A water bottle with sprayer is handy on old mine dumps.
12. Investigate the ground around old dumps and old mines.

Exquisitely detailed crinoids come from near Elkader in Cretaceous outcrops. Photo by Don Goodart, courtesy of South Dakota Museum of Geology.

Ammonites from western South Dakota are beautifully preserved. Photo courtesy of Black Hills Institute of Geological Research.

FOSSILS

Western South Dakota is a world-class fossil locality. Vertebrate, invertebrate and plant fossils from here grace most major earth science museums around the world. Few places expose more species of fossils than the White River badlands. The badlands, magnificently sculptured by wind, ice, sand and water, were once a part of a vast sea which stretched from what is now the Gulf of Mexico to the Arctic circle.

All collecting is forbidden in the Badlands National Park, however it is worth a stop here to see the small museum and to follow the marked "Fossil Trail". Notice that not all badlands are spelled with a capital "B". The capital letter is used only for the Park. The word badlands is a geological term referring to a precipitous terrain, carved in differential sediments by rapid runoff of heavy rains or melting snows.

The government has new regulations about collecting fossils. Get these through the Bureau of Land Management, Department of the Interior or from the National Grasslands, Department of Agriculture. Invertebrate fossils and plant fossils may be collected on Public Land if they are not sold. There are some restrictions about vertebrate fossils. While you can pick up an isolated tooth or piece of bone, if you find a skull or an articulated fossil, you should cover it up, mark the spot, and notify professional paleontologists.

South Dakota fossils are displayed at the Museum of Geology in Rapid City. They include the Tiatanothere, the Oreodont, sabre-toothed cats, ancestral camels and rhinoceroses, primitive horses, crocodiles, elephants, and giant turtles.

Among the most desirable invertebrate fossils are the marine fossils of the Pierre Shale formation, which is dominant in much of the western part of the state. The Cretaceous fossils are found in huge "mud ball" concretions along the Cheyenne River, and other rivers and streams. Best-known of these fossils are the beautifully preserved ammonites, scaphites, and baculites. It is also possible to find a large iridescent clam or an elegant many-chambered nautilus. Intricate details, large size, and smooth luster are the hallmarks of the best specimens.

Look for fossil-bearing concretions along the Bad River in Stanley County, the Moreau River in Dewey County, and all along the Cheyenne River, and in road material quarries and road cuts. Some private land owners will give permission to hunt.

When you find a concretion, open it carefully from the top with a maul, somewhat like taking the lid off a box. Calcite and barite crystals are found in these concretions in certain locations. Parts of the concretion are often highly fluorescent.

The Black Hills Institute of Geological Research in Hill City, South Dakota has an exceptional display of vertebrate and invertebrate fossils from South Dakota and other places. This group is known for their expertise at preserving and restoring fossils. Their cephalopods are among the most beautiful ever seen and their carefully prepared duckbill dinosaurs from western South Dakota are highlights of several major museums.

Plant fossils are abundant throughout the badlands and Black Hills. Water-rolled rocks, upon close examination, may have remarkable cell patterns, indicating that they were once part of the vegetation of a warm and humid area. Among the rare plants are tempskya, sequoia, and cycad.

Cycads are known as the earliest flowering plant, an ancestor of both palm and pine. The cycads were low plants with a crown of stiff but graceful fronds, and bases shaped like king-sized pineapples.

Characteristic of the cycad wood, is a pattern of bold diamond shapes. The field collector of today is more likely to come upon broken and worn pieces of cycad, than the large intact specimens once found in abundance in Fall River County. Any of the rock beds south and east of Hot Springs may have pieces of cycad and other exotic wood. Among these are the Ardmore and Oelrichs rock beds.

Another fossil plant from near Hot Springs is "tufa". Tufa is petrified moss, water cress, or an assortment of plants preserved by the deposition of calcium or silica by hot water. Some of the gray and buff specimens show delicate details. Fragile-appearing specimens may look like stone carvings of graceful ferns, while coarser material may resemble stone cattails and reeds.

The tufa may be found on the banks of Cascade Creek and Fall River. One location is near the ghost town of Cascade, an early day spa, south of Hot Springs. Take S.D. 71 south of Hot Springs 9 miles. On the right (northwest)

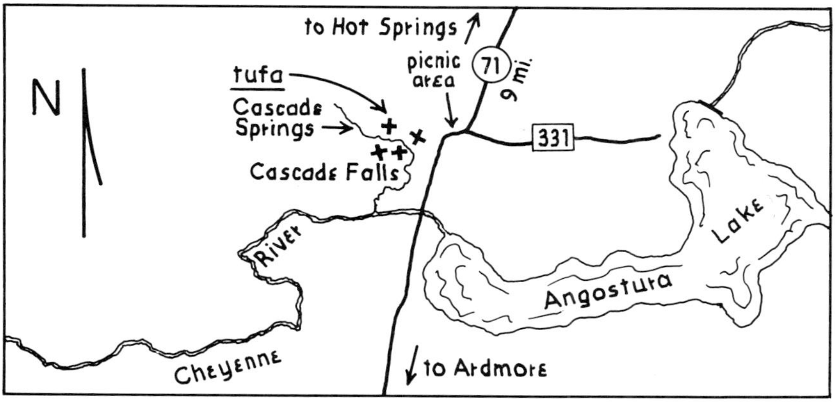

Tufa south of Hot Springs on Cascade Creek

is a government-operated picnic grounds. Park here and follow the creek. Shortly south of here the tufa is found in the stream banks.

In the city of Hot Springs is an extraordinary fossil exhibit, the Mammoth Site, where the remains of 40 Columbian mammoths are being excavated, preserved, and shown in place. Visitors can view the incredible tusks and massive skulls in the spot where they perished 26,000 years ago. Designated as a National Landmark, the fossil site is covered by an impressive modern building. This is the largest mammoth dig in the world.

Eastern South Dakota

As stated earlier, the Missouri River slices South Dakota in half with a series of large dams and lakes. The hilly rangelands of the west, marked by rugged canyons, buttes, and badlands, abruptly give way to the rich treeless farmlands of the eastern part of the state.

The gems and minerals and fossils of the eastern half are as different as the terrain. Except for the state parks, all of the land is privately owned. This is a glaciated area, so the gravel pits will have many rocks and fossils which originated farther north and were left by the glaciers.

Lake Superior type banded agates are found along many of the streams and glacial lakes in the extreme eastern part of the state. Big Stone lake, Pickeral Lake, and Big Stone Lake are some of the places nice agates have been found. Look also in gravel pits near Watertown, Milbank, Waubay, and Brookings. Some of the things which occur in the gravels are petrified wood, fossil corals, chalcedony, quartz, jasper, feldspar, and chert.

Mammoth bones and teeth and other Pleistocene fossils have been found near Aberdeen, Gettysburg, Mobridge, Huron and Big Stone City.

Club members of the Corn Palace Club in Mitchell have found thousands of

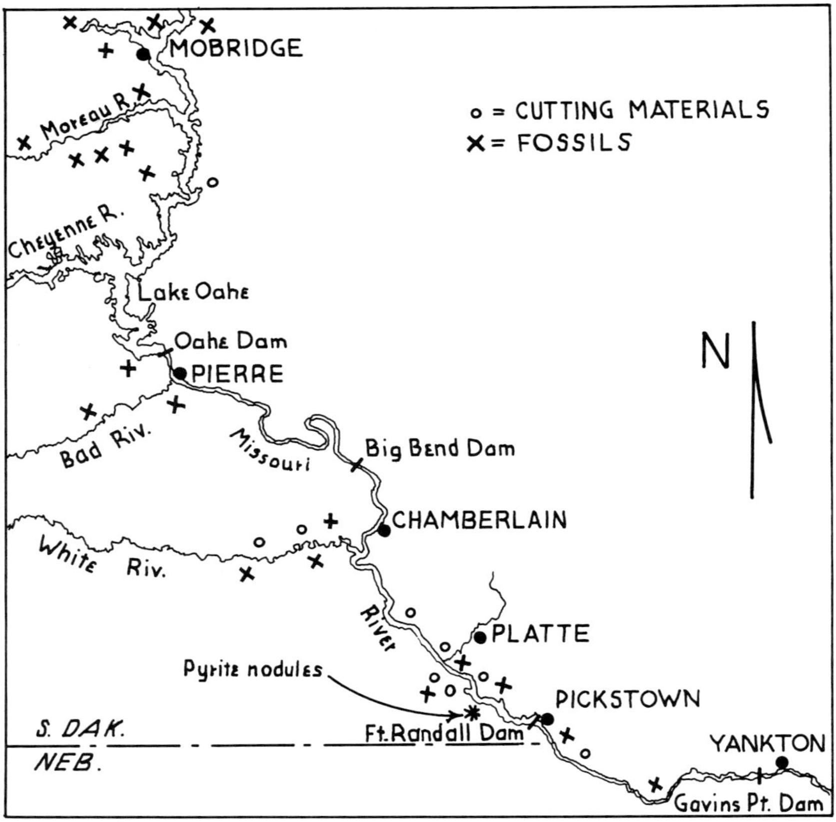

Rocks and fossils along the Missouri River

fossil shark's teeth in and along Davidson County's Firesteel Creek. Fossil fish occur near Ree Heights, Hand County.

Fossils are found in some of the limestone and chalk outcrops in bluffs along the river from Fort Thompson to Yankton.

Conglomerate is found along Split Rock Creek from Sherman to Garretson.

Fossils and various kinds of cutting materials may be found all along the Missouri River throughout the state. The fossils are marine invertebrates, and sometimes fish fossils, and are embedded in concretions or rocky strata. People often rent boats at marinas and go along the edge of the river to look for fossiliferous banks. Some of the places to look for fossils are near Mobridge, Pierre, Chamberlain, Platte, Pickstown, and Yankton.

Also along the river are many water-worn rocks which look quite ordinary, but have proven to be quite handsome when cut and polished. A judge who lived at Platte brought home and sawed up all colors of river rock and found some to resemble jade, jasper, unakite, epidote, Sioux quartzite, flint, mozarkite and many other better known materials. Such rocks are large enough for bookends, pen sets, paperweights, plaques, and other decorative and practical items. An attractive porphyry and jig-saw puzzle breccia are among the interesting finds. Rock piles in the corners of the farmers fields have the same types of rocks as those found along the river.

In the clays of Aurora and Spink Counties, ornate selenite roses occur. Watch any excavations in the heavy clays for the silvery glint of selenite.

The red Sioux quartzite of the Sioux Falls area is a classic and durable building and decorator's stone. Massive amounts of the hard pink to red stone are found over wide areas and have been quarried for a century and provided ornamental material for many purposes. Some of the quarries and outcrops are at Sioux Falls, Palisades Park, Dell Rapids, Spencer and Alexandria. In these same areas, catlinite, the revered "pipestone" of the Sioux occurs.

Near Milbank in Grant County, there is an extensive deposit of deep red granite. This granite has been quarried for many years and is sold for monuments, memorials, and ornamental building material all across the nation. Called "Mahogany" granite, the rock owes its color to the bright feldspar component.

Pearls have been found in mussels of the Vermillion River in Clay County. The lustrous fresh water pearls occur in the native mussels of unpolluted stretches of running water.

Permission to collect is required for all eastern South Dakota localities.

Where to learn more

The Museum of Geology at the South Dakota School of Mines in Rapid City is world famous. There are also museums with earth science materials in Aberdeen, Sioux Falls, Mitchell, Murdo, and at the University of South Dakota in Vermillion. South Dakota clubs sponsor field trips in connection with their shows. The annual shows of the Rapid City, Sioux Falls, Mitchell and Hot Springs clubs are well-attended and have educational displays of local materials. A smaller educational show alternates between Winner and Mission. The South Dakota State show is held in a different city each year.

There are more rock shops in South Dakota, for its size, than in any other midwestern state. Most of the shops have native materials, and some of the owners will give directions to field collectors. Shops which have been cooperative with collectors are in Sioux Falls, Okaton, Rapid City, and Hot Springs.

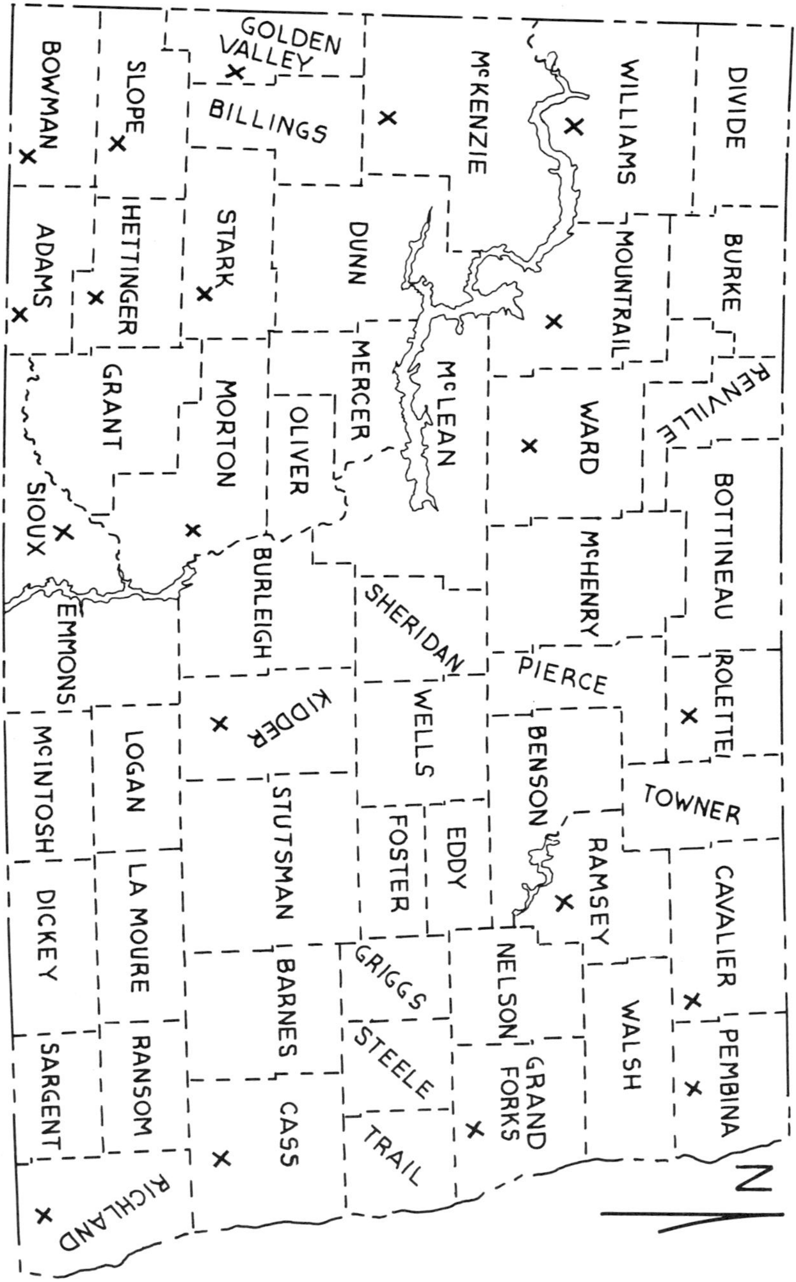

Xs show locations for fossils, gems and minerals in North Dakota

NORTH DAKOTA

North Dakota is a leader in lignite production and has vast reserves of this soft coal. The state also has oil, sand, gravel, clay and shale. Western North Dakota is a land of rolling hills, abrupt scenic buttes, and eerie sculptured badlands. Theodore Roosevelt National Park near Medora has prominent erosional features in rich and varied colors, and areas of cedars and pines. The state has become a mecca for tourists since Theodore Roosevelt became so enthusiastic about its natural attractions that he made it his home. While the western part of the state is rugged prairie and steep canyons, the eastern part of the state tends to be more level, with fertile agricultural land and glacial lakes. Damming the Missouri River has given the state many miles of lake shore.

Western and Central North Dakota

Moss agate is found along the Little Missouri River, and on the hillsides and eroded places in the southwest portion of the state. Identical to the famous Montana moss agates, the nodules are found near federal, state and county roads. One place to look is in the Williston area. Near this oil-boom city, the Yellowstone River, home of the Montana agates, empties into the Missouri River, helping to form one of the mightiest rivers in the nation.

The ever popular lapidary material may be difficult to find until the collector gets used to its appearance in the rough, which may have the translucency disguised by a white paint-like coating. The agates are present in alluvial gravels which do not share this unique coating.

The Montana-type moss agate is a favorite with lapidaries everywhere. The pyrolusite dendrites, resembling moss, trees, flowers, and sunbursts are usually deep brown or black in color, and often lighter in color towards the edges. The chalcedony nodules which include the scenic dendrites are translucent and colorless, whitish, or slightly grayish.

There are also brownish or redish arcs or partial bands in many of the agates. A very few when cut thin and backlighted will show rainbow colors and are classed as iris agate.

Some agates have trees or other mossy patterns with red tops, and are appropriately called "red tops". Such agates may also have streaks and clouds of red-brown.

The potato-shaped nodules are often covered with a white coat of opaque chalcedony, altered by weathering. The nodules are water-worn. The chalcedony is hard, uniform, and possessed of a sharp conchoidal fracture. The sizes range from small plum size to large cantaloupe size, although the most common finds are 3 to 5 inches in diameter.

Sometimes the dendrites are along one plane forming a scenic pattern. Sometimes the spots are very small and give a gray polka dot effect. Other spots may follow the concentric bands of the agate's growth.

Montana-type agate is easy to cut and polish and has been used for popular jewelry for many years. The highways in North Dakota which are access routes to Montana-type agate localities are U.S.Highways 2, 94, and 12, all entering North Dakota from a prime agate area of Montana.

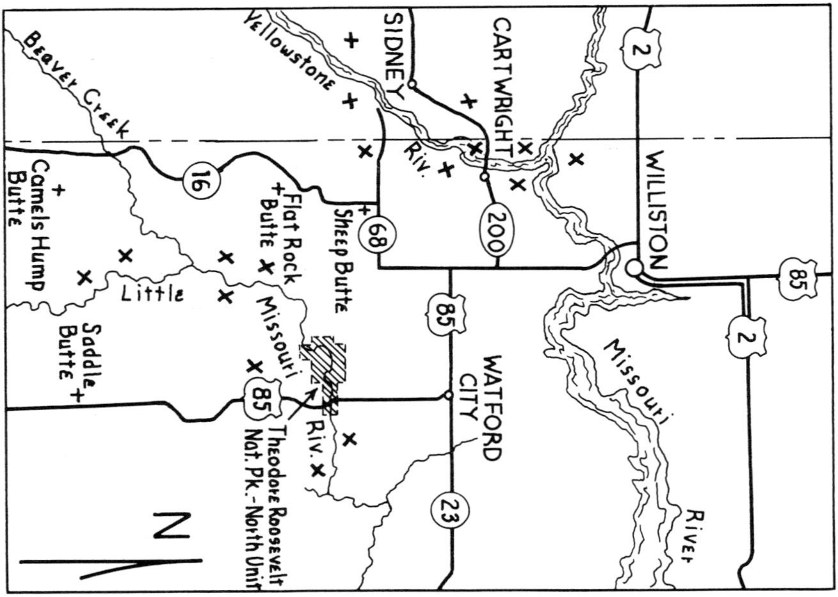

Xs indicate agates, jasper and petrified wood along the Yellowstone and Little Missouri Rivers

The state routes near the agate areas are 200, 68, and 16. Agates may be found near Sheep Butte, Flat Rock Butte, Camels Hump Butte, Sentinal Butte, Chimney Butte, Bullion Butte, Pretty Butte and Rattlesnake Butte.

Other quartz family gems are also found in this part of the state. Permission is always required on private land. Gravel pits, grassy fields, stream banks and bars, and eroded areas all offer good collecting for chalcedony, picture agate, tube agate, iris agate, carnelian, sardonyx, onyx-banded agate, bloodstone, concentric banded and fortification agate, and multi-colored jaspers. There is also petrified wood.

Cryptocrystalline quartz gems are found along the Little Muddy in Williams County, featuring some very showy petrified wood. McKenzie County, near Cartwright has been a productive agate area. Other sites are the alluvial gravels near Dore and Buford and along Beaver Creek and east of Trotters and Grassy Buttes.

In Billings County Cretaceous vertebrate fossils have been found on Andrews Creek and Bacon Creek and near White Butte and Yule. Uranium has been mined in Billings, Mercer and Golden Valley Counties. Rich ore was found near Sentinal Butte.

Agates, jasper, petrified wood and fossils are found along the Little Missouri River in Slope County north and west of Amidon. White Butte, the highest point of North Dakota at 3,506 feet is south of Amidon.

Other cutting materials are found in Bowman County along the Little Missouri south of Marmarth. Chances are good here for petrified wood, chalcedony, agates, jasper and fossils.

PETRIFIED
PARK
WELCOME

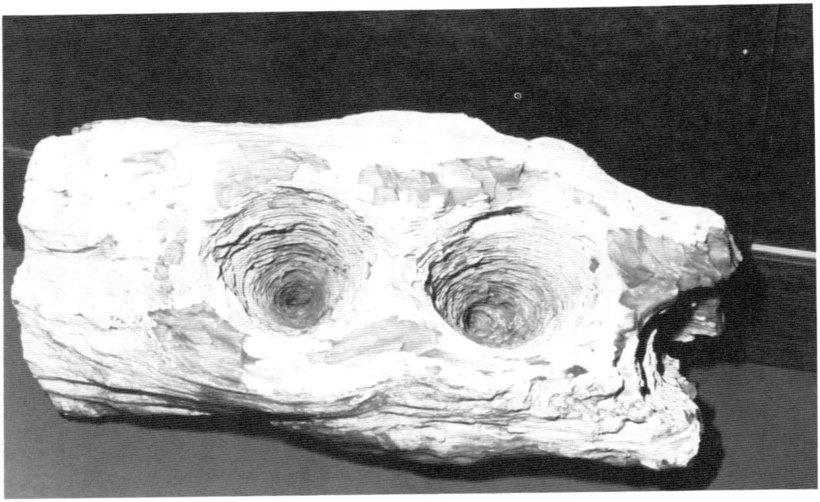

Well preserved and finely detailed petrified wood is found near Hettinger, North Dakota. William Burish Collection. Photo by Central Dakota Glen and Mineral Society

Hints For Collecting Gem Materials

Following are some suggestions for collecting gem materials.

1. Walk back and forth looking at the rocks with the sun in front of you and then behind you. Agates and chalcedony are translucent with the light shining through them. Patterns show off better with the sun behind you.
2. Look for gemstones after rains if possible. Moisture makes the patterns and color stand out.
3. Learn to check rocks in the field. A small chip can be knocked off from the edge with a prospector's pick by holding the rock firmly with one hand and striking it quickly at the edge with a hard blow. Let your wrist give with the blow.
4. If a rock is already broken, a conchoidal fracture is often a sign it is polishable.
5. Look in streams and along lakes, and also on the grassy hillsides and dirt banks along the lakes and streams.
6. If a rock is of good color but badly fractured, remember your tumbler.
7. Agates may have oxidized coatings. They may also have thumb print depressions or pockmarks.
8. Make a record on the spot of any find you make which is strange or unusual in any way.
9. A spray bottle of water will help determine the characteristics of individual gem materials.
10. The collecting of petrified wood is regulated. You may not take more than 25 pounds in one day or 250 pounds for a year.

A good region to look for polishable materials is in Hettinger County near Mott. The route between Mott and Richardton to the north through the county has numerous rock beds. Exotic petrified wood is found at Camel Butte north of Mott. Jasper, chalcedony and several kinds of agate occur in fields, quarries, washes, and roadsides.

Dickinson is a Stark County community located near the center of the North Dakota section which is southwest of the big curve made by the Missouri River. Mammalian fossils have been found 12 miles south of Dickinson. Among the fossils were many extinct species such as bones of the ancestral horse. This is an eroded badlands type area of Oligocene "White River" beds.

A record find in North Dakota was a petrified tree 120 feet long discovered near Dickinson during the construction of Interstate 91 . The tree was 10 feet in diameter at the base. The silicified wood is brown, white, black, gray, tan and slate blue.

Some of the petrified wood of North Dakota is only partially petrified, and indeed a match will burn some of it. Other specimens are well-agatized and sometimes partially opalized. Strange specimens appear to be turning to lignite.

The banks, terraces, tributaries and gullies of the Cannonball River have a wide variety of gem materials and plant fossils. The Cannonball River rises in Slope County north of Amidon and flows through Hettinger County, Grant County, Sioux County and Morton County, entering the Missouri River at the town of Cannonball in Morton County. Near the town of New England in Hettinger County on the Cannonball, petrified wood and other fossils are found.

Xs indicate collecting areas south and east of Dickinson

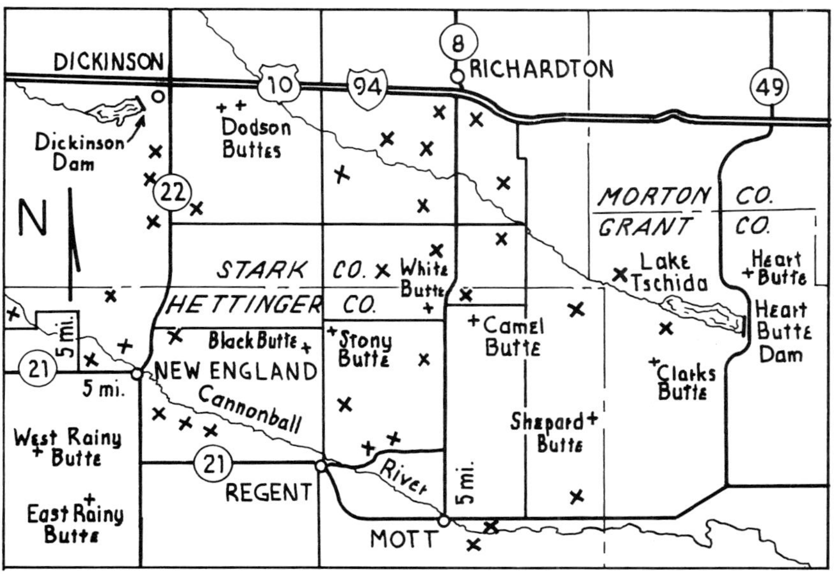

Gullies, steep banks, gravel bars, and terraces along the Cannonball and Cedar Rivers in Adams and Sioux Counties have been the scenes of important finds in the past.

Petrified wood and well-silicified wood seems to be the best-known and most available of the many interesting materials of western North Dakota. Some of the wood has good color, excellent detail, and takes a glassy polish. Sometimes very large pieces are found. Most of the wood is easily identifiable.

Much of the North Dakota petrified wood is float material, but in some badlands areas upright petrified stumps have been found standing. (One of the most persistent rumors among North Dakota Old Timers, is that they remember a particular tree from their youth, 80 or so years ago, and return to find it partially petrified.) Though such reports sound imaginative, to say the least, some of the North Dakota wood is poorly petrified and not silicified.

Large quantities of petrified wood, huge logs and trunk sections, and limbs, were taken from abundant North Dakota sites to the Petrified Wood Park, a city park and landmark in Lemmon, South Dakota, just over the border.

Any collector hunting petrified wood in the river breaks of North Dakota, South Dakota or Nebraska, should be aware that where there is petrified wood there could also be petrified bones, or seeds of these ancient fossil plants. In hunting for small seeds of prehistoric trees, gem screens are a valuable aid. Where there is a concentration of petrified wood, shovel the gravel onto nested screens. The seeds may be as small as hackberry seeds or as large as cones, but their origin is usually obvious as the more irregular pebbles are discarded.

The most interesting petrified wood is that which has the finest details of grain and growth rings, or that which has knots, twigs, bark, or some other distinguishing feature. Sometimes the North Dakota wood which is not of cutting quality turns out to be extremely informative to paleobotanists. Much of the wood is cuttable however, and it too tells paleobotanists a lot about the flora and climate of this region in the distant past.

The Cannonball River breaks in Grant County have yielded examples of ancient forests which are of great value to science. The late John Bunting of Mandan found silicified sequoia cones, and specimens of osmundite, tempskya and katsura in this region. Since then other fine specimens have been found by knowledgeable collectors south of Leith and Brisbane and near Shields.

Bluffs near Brelen on the Cannonball in Morton County have fossil exposures at the base. Fossil shells are found in sandstone concretions along the river. Fossils have also been found at the Sweet Briar Dam and Reservoir.

The State Fossil of North Dakota is unique in that it combines both the animal and plant kingdom in the third kingdom, mineral. It is teredo-bored petrified wood. This unique wood is ancient driftwood into which mollusks bored passages. Although most of these teredos, or shipworms, were small, some were as large as 16 to 18 inches in length. These particular pelecypods lived in seas of the dim past; similar species are found today in pilings, piers, boathouses, and wooden boats.

Teredo-bored wood is a mark of the Cannonball formation on both sides of the Missouri River in the vicinity of Mandan and Bismark.

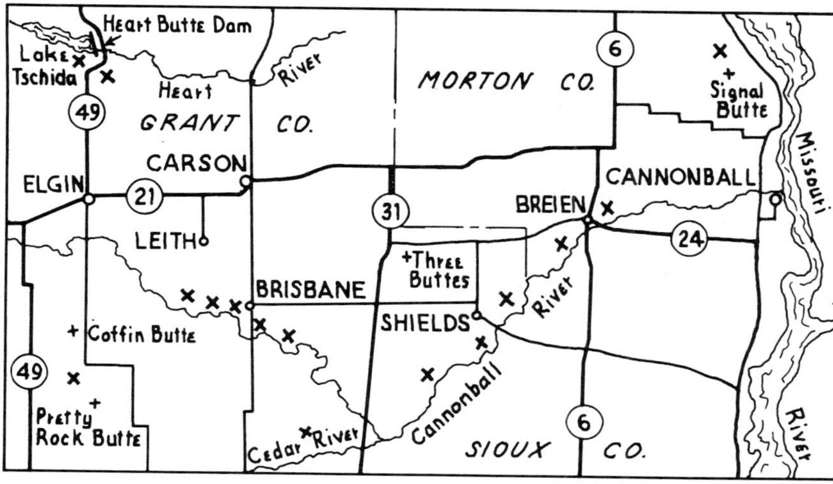

Xs mark fossils and gem material locations along the Cannonball River

The overall appearance of the teredo-bored wood is of a polka dot pattern gone astray. Various minerals have filled the borings. Some of the fillings are of gem material such as quartz, chalcedony or opal, even occasionally precious opal. Sometimes the borings are cavities, which may be lined with drusy quartz or amethyst. Some of the wood is highly agatized, while other specimens are more interesting as cabinet pieces rather than cutting material. Some of the teredo-bored wood has been found to be sequoia, and other specimens are deciduous wood.

Teredo-bored petrified wood has been found in gravel pits, and along rivers, and in eroded sediments in the southwestern part of the state. Gravels near the Missouri and Heart Rivers in the Mandan area have yielded fine specimens of this fascinating material for many years.

Pyrite and marcasite have been found in the strip mines of the lignite beds of western North Dakota. Pyrite nodules are found in some of the stream gravels. Plant fossils of several species are associated with the lignite. In some spots underground lignite fires of the past have turned the surface clay to a vivid red material called scoria.

In Burleigh County, 1.1 mile south of Moffat, superb fossil crabs were found.

In Emmons County, Pierre shale outcrops along Beaver Creek have ammonites, scaphites, baculites and clams of the Cretaceous. Look for more cretaceous fossils 4 miles north of Linton.

Fossil-bearing concretions from the Pierre formation sometimes have calcite crystals, or aragonite, or velvety siderite. Selenite crystals are found near the saline lakes of Montrail County and in many clay beds in the state. Glauberite is a mineral found near Stanley in Montrail County, and also sometimes thenardite and halite. There are gravel quarries south of Minot in Ward County.

Gravel pits along the Missouri River near the state capital Bismark, have agates, jasper, petrified wood, and other materials of interest to lapidaries and field collectors.

Eastern North Dakota

A line through the center of North Dakota would be shortly west of the 100th meridian. Rich farm lands and rangelands occupy much of the eastern half of the state.

Gravel deposits are mined at various locations in the warm months. Some of these are near Tappen and Crystal Springs near Interstate 94 in Kidder County. Look for jasper, chalcedony, agates, petrified wood, and fossils in the quarries.

The Turtle Mountains of the north central portion of the plateau region have manganese reserves. Gravel has been produced in Rolette County. Most of North Dakota has been covered by glaciers several times, so it is possible to find a wide variety of hard stone in the glacial gravels around many of the lakes.

The eastern part of the state was covered by the immense Lake Agassiz, which also covered part of eastern South Dakota. There are gravels of interest to collectors in the Altamont moraine. The James River and the Sheyenne River have been hunted with varying success.

An area in the northeastern corner of the state on the Tongue River called Concrete, has an outcrop of fine blue-gray limestone, where fossils and sometimes crystals are found. Near Lisbon in Ramsey County on the Sheyenne River some striking teredo-bored petrified wood was discovered. It was jet black with golden colored chalcedony fillings. Other quartz gems are found here too.

Near Wahpeton, Fargo, and Grand Forks, Lake Superior type banded agates are found in the gravels, along with carnelian, sardonyx and chalcedony.

On the eastern border of the state is the Red River, one of the few in the United States to flow north. Agates and petrified wood are found in the Red River Valley. One location is the Erie Dam southwest of Erie in Cass County.

Little is known by most hobbyists about the gem materials of North Dakota. The state is sparsely populated and most of the land is privately owned. Since many of the gem materials are drift or float, having been deposited by water or glaciers, there are few spots where the collecting is always good. The collector's chances of exceptional fossil finds are somewhat better than finding excellent gems or minerals. For a collector with time and a knowledge of geology, North Dakota could be a frontier. Wherever there are excavations, construction projects or land clearings, unusual and strange rocks and fossils may turn up.

North Dakota has earth science museums at Wahpeton, at the State Science School, at the University at Grand Forks and at the College in Fargo. The Broste museum of gemstone spheres is at Parchall.

Longitudinal borings and cross sections of teredo borings are seen in this prized North Dakota petrified wood.

The state has several active gem and mineral societies and also a paleontological society which meets in Bismark. Club members helped make the teredo-bored petrified wood the State Fossil. An excellent place to see what North Dakota has to offer is the annual show of the Central Dakota Gem and Mineral Society.

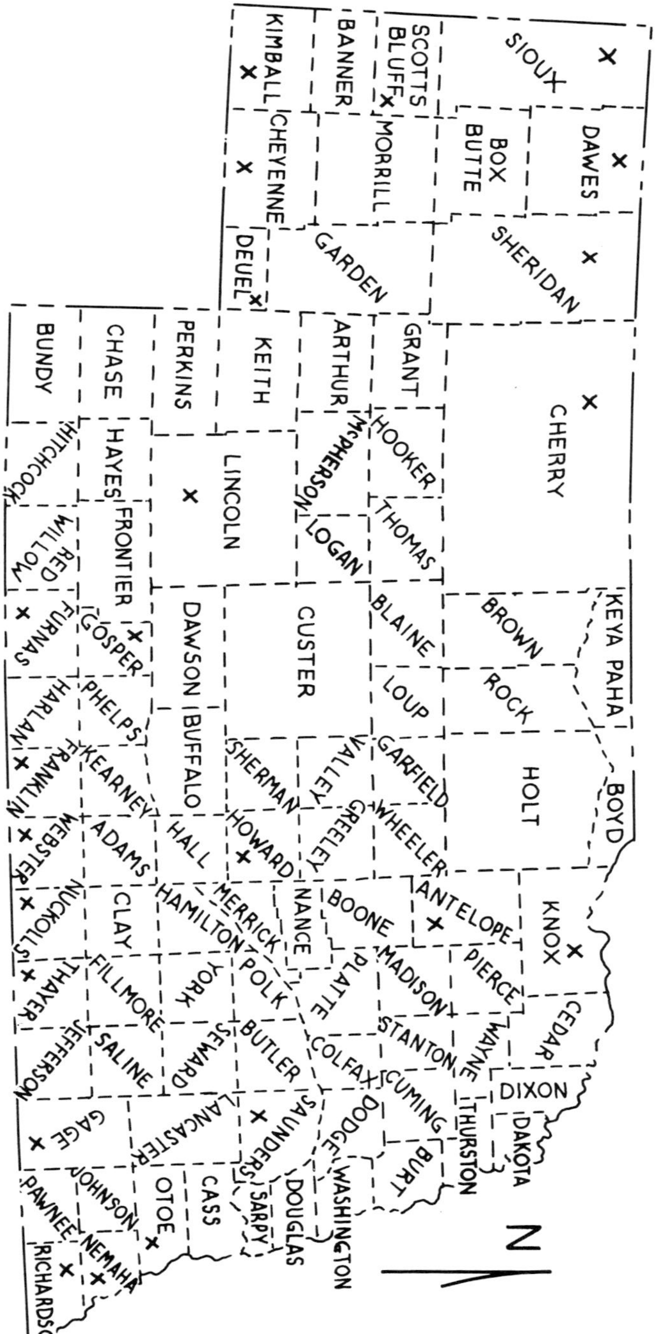

Xs. show gem, mineral and fossil locations in Nebraska

NEBRASKA

Nebraska is a central state with a wide variation of terrain from eerie badlands, and stark buttes, to pine-covered hills, rolling sand hills, and fertile agricultural valleys. Except for the National Grasslands, the Nebraska National Forest, and the State Parks, the land is privately owned. Fossils may be the most collectable of Nebraska materials, nevertheless colorful agates, high-quality petrified wood, and other lapidary materials are found throughout the state. Excellent crystals have been found in concretions and quarries. Broad rivers bisect the state from west to east, and the wide Missouri and its bluffs mark the eastern border.

Western Nebraska

The National Grasslands of western Nebraska furnish collectors with some excellent hunting for quartz gem materials. In Sioux and Dawes Counties the rock beds of the National Grasslands are like the alluvial rock beds of western South Dakota. The materials to be found include brilliant and boldly patterned jaspers, several pastel colors of chalcedony, bright well-agatized wood, carnelian, Montana-type agate, Fairburn agates, prairie agates, quartz after gypsum pseudomorphs, barite nodules, and other numerous beauties and oddities of the mineral world.

Well-known badlands-type agate beds in the National Grasslands are 20 miles north of Crawford, where a rough and eroded terrain defines the breaks of the White River. The Highway is marked with numbers 2 and 71. Toadstool Park is a popular campground near excellent collecting areas, convenient to agate beds near Ardmore and Orella along the northern border of the state.

Agate beds and fossil locations in Northern Nebraska

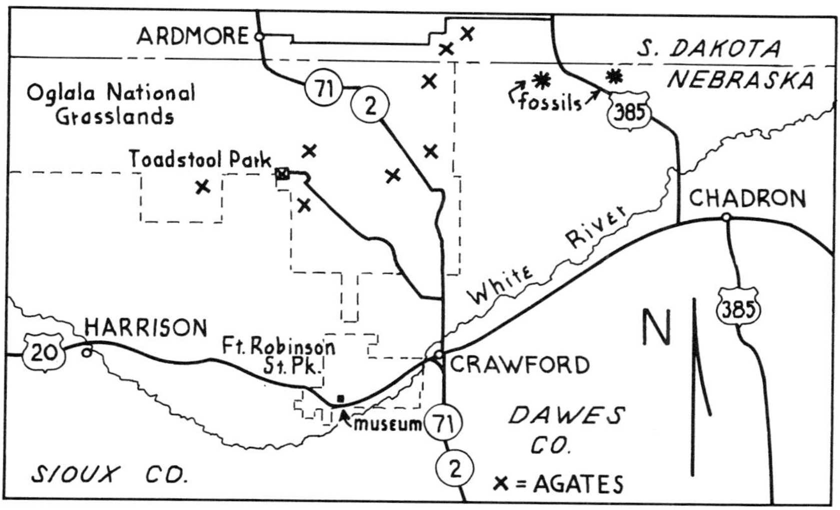

The route to the campground turns west 5 miles north of Crawford and then heads northwest. The route is marked. Not only are there plentiful rocks in the vicinity, but there are also some very fine fossils. (Most vertebrate fossil collecting is forbidden on Federal Land.)

Other areas are 1.5 miles east of the old site of Orella, 18 miles north of Crawford and stretching into Dawes County. Another route to the National Grasslands is north of Harrison, a community 27 miles west of Crawford. Look in road cuts and grassy areas besides the rock beds. The Nebraska State Gem, blue agate, and the State Rock, Prairie agate both come from the Orella area, and the Grasslands north of Crawford. The blue agate or chalcedony, is found in-place in seams and fillings, in the siltrock of the Chadron formation. Scarce now on the surface, except for scattered small pieces, agates have been found in larger sizes by digging. The Chamber of Commerce at Crawford has a map of the important collecting areas north of this city, and can give current information about access routes.

Gypsum roses replaced by chalcedony are among the most fascinating objects of these northern Nebraska rock beds. Much of the chalcedony of the rock beds is fluorescent. Colors under short wave lamps will be vibrant greens, pinks, and deep reds. Sometimes the chalcedony of the region has swirling patterns of black, brown, white, or cream. Silky botryoidal cavities occur in some pieces of chalcedony.

Jasper (or chert) is the most abundant lapidary material here. "Prairie agates" are usually more jasper than agate. The colors of the bands are pink, red, yellow, orange, gray, and blue. However the banding is in the form of wavy swirls, and is not parallel and sharply distinct as in the Fairburn. Solid color jasper nodules cover the spectrum. Those which appear black on the surface are often vivid red within. Some are translucent and qualify as agate. There are also pieces of jasper conglomerate with bright reds and yellows, and colorful breccia cemented with chalcedony. Marine fossils replaced by jasper, or jasper with fossil imprints are other collectables. Many of these items should not be chipped, and pieces which look "borderline" should be chipped only judiciously.

The petrified wood of the region is exotic, well-agatized, and with fine details of grain preservation. Some of the wood is colored by warm hues of red, yellow and gold. Examine the wood closely. Once in a while it will have a pattern so unusual that it will send the finder to fossil plant reference books. It may be cycad, tempskya, osmundite, pteridosperm, or "teredo wood". Agatized bog found here consists of bits and pieces of fossil plant materials replaced and cemented together by agate or chalcedony. Novice collectors sometimes have trouble recognizing the petrified wood, as it is water-worn, as are the other materials. The wood pieces are more irregular in shape than the jaspers, and the structure appears to be formed of longitudinal fibers. Pieces of wood sometimes have knots, or places where twigs have been broken off.

In spite of all the desirable materials found in these rock beds, the most hunted of all is the Fairburn agate. The Nebraska Fairburns are similar to the South Dakota agates, but it has not been determined if they are from the same geological origin. The prize agates can be found in the Oligocene gravels from the Wyoming border on the west and the South Dakota border on the

An agate found in western Nebraska, looking decidedly drab from the outside, explodes with shimmering brilliance when cut and polished. Courtesy of Nebraska Geological Survey.

north to Sheridan County in the east and the Scottsbluff area on the south. The National Grasslands constitute the best areas for these scarce agates. Hunting is allowed in National Grasslands without special permission, however loopholes in the regulations have allowed some ranchers to fence off some areas so there is no access. There is a National Grasslands office at Chadron.

Many agates have been found on private lands also, so wherever you can get permission to hunt, look in eroded areas and hillsides with sparse vegetation.

The Nebraska Fairburns are bright fortification agates with finely detailed bands of contrasting colors. The patterns tend to occur more often in matrix than in the South Dakota agates. The matrix is most often yellowish or reddish brown and unpatterned, however it is usually polishable, unless the jaspery material is of poor quality. The agates which have weathered out to solid, smooth agate nodules are as beautiful as they are hard to find. However, the chances of finding one may be just as good as in South Dakota.

Northwest of Hay Springs there are large septarian concretions which have crystals of calcite, barite, celestite, and sometimes such marine fossils as clams or ammonites. Concretions are also found 20 miles north of Chadron.

Sand barite nodules are sometimes found in badlands areas. Tan in color, they look somewhat like flattened walnuts. Selenite crystals are sometimes found on the surface in dry washes.

Colorful cutting materials are found along the rivers and in the sand bars and on the terraces in the Nebraska "panhandle". The rivers are the White, the Niobrara, the North Platte, the South Platte, and Lodgepole Creek, which empties into the South Platte in Colorado near the corner of the panhandle.

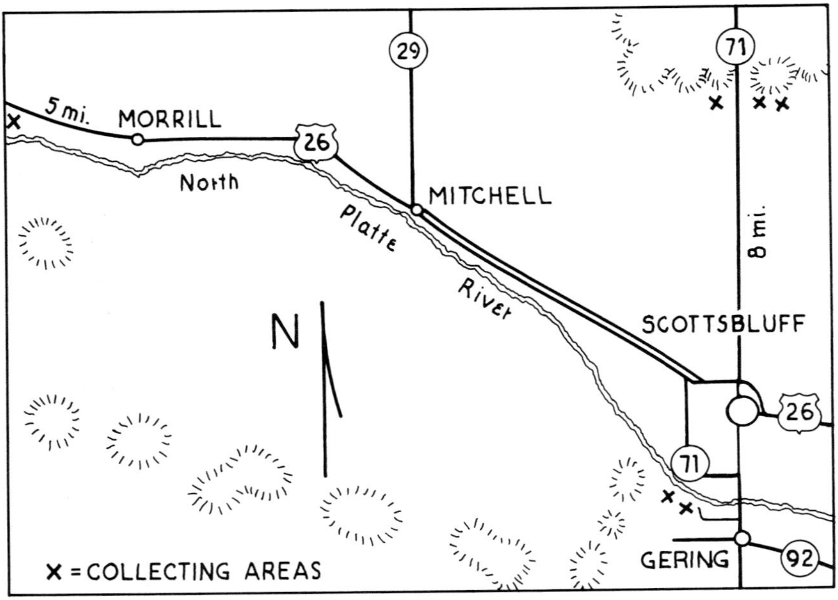

Concretions, cutting materials and fossils in Scottsbluff area

The materials to be found are moss agate, banded agate, chalcedony, prairie agate, jasper, carnelian, sardonyx, onyx-banded agate, flint, chert, and petrified wood. Some of the petrified wood is brightly colored and well silicified.

Always ask permission before collecting along rivers or in the sand and gravel pits along the rivers. The best time to hunt along rivers is when waters have receded and are calm. River currents can sometimes be treacherous, so stay along the banks. Gravel bars on streams can be excellent collecting sites. Look along the tributaries of the North Platte and South Platte Rivers for the many quartz materials which have weathered the ages.

There are no fee areas in western Nebraska, but members of the Plains Rock and Mineral Club at Bushnell are able to advise visitors. A particularly good stretch for petrified wood is between Bushnell and Chappell.

A marvelous variety of quartz gems occurs along the North Platte River in western Nebraska. This major river enters Kimball County Nebraska from Wyoming, the jade state, so a few pieces of float jade have been found in Scottsbluff and Morrill Counties. Also pieces of facetable smoky quartz and polishable microcline and labradorite have added to the variety. Many of the river specimens are of high quality.

Cutting materials and fossils are also found 8 miles north of Scottbluff both east and west of Highway 71. There are also concretions here. More fossils, agates and petrified wood are found southwest of Scottsbluff near Gering on the south side of the North Platte.

The Mormon Trail and the Oregon Trail follow the wide river, so state routes

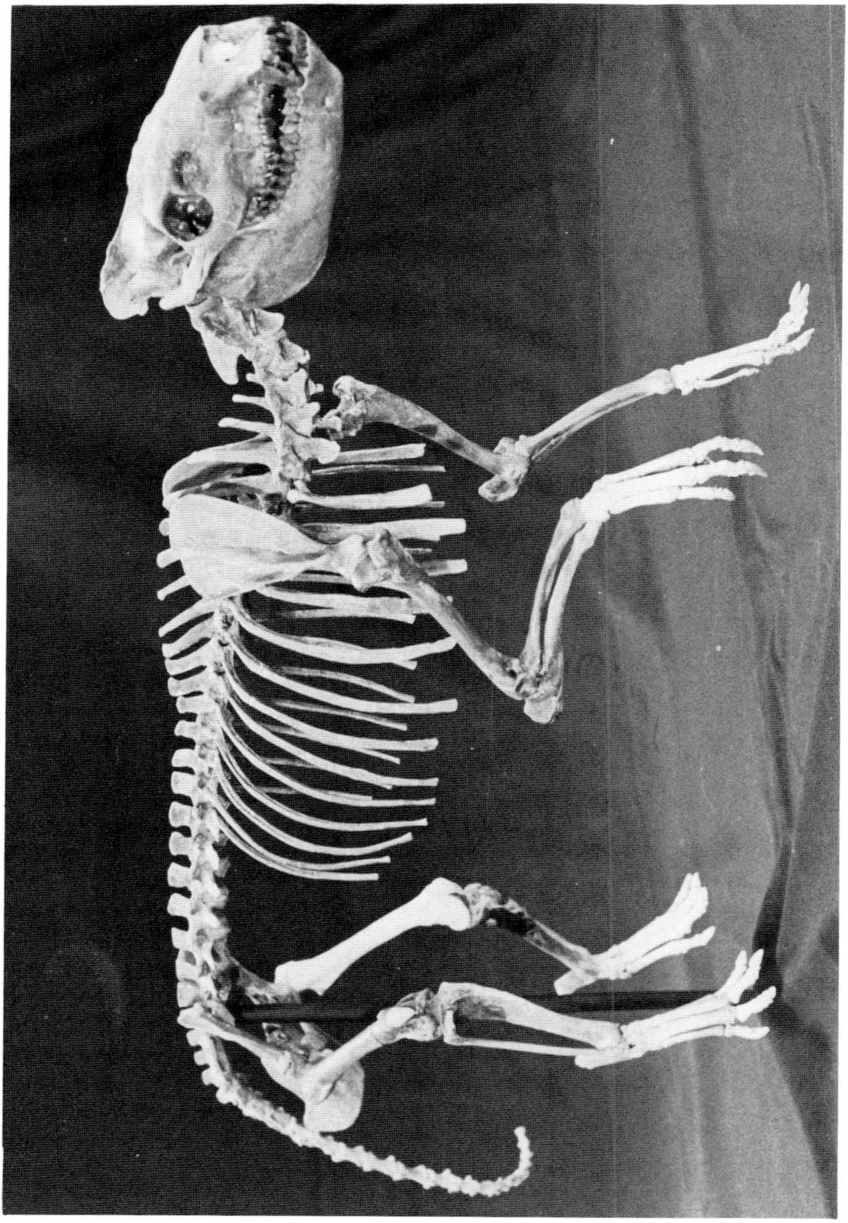

This Oligocene Oreodon skeleton comes from Sioux County, Nebraska. Photo courtesy Black Hills Institute of Geological Research.

Fossil marine invertebrates of the Cretaceous Period are found in Dixon County in northern Nebraska.

Lake Superior agates are to be found along rivers and creeks in eastern Nebraska.

which follow the river,and are the present access routes for collectors, are of historical interest as well as geological.

In the hills near Chappell in Deuel County, bordering Colorado, there have been significant finds of petrified wood, chalcedony, agate and jasper. Banded jasper is found in Morrill County near Bayard. In this same county is an old location for moss agate and moss opal near Angora.

The Niobrara in Cherry County provides another opportunity for the collector to find petrified wood. The wood is sometimes opalized and large pieces are sometimes found. Fossil leaf imprints and remnants of vertebrate fossils also occur in this county. Other rivers in this Sand Hill area have gems, fossil vertebrates and invertebrates, and fossil plants. The other waterways of the Sand Hills are the Snake River, the North Loup River, Spring Creek and Minnechaduza Creek. There are also dozens of spring fed lakes.

It is hard to imagine such an immense dune area in the middle of the nation. This is sparsely populated ranching country, often desolate and treeless, except for the cottonwoods near the water. But near Valentine there is a ponderosa pine forest with scenic deep canyons.

Much of the wood found near Valentine is opalized and extremely brittle. Huge logs have been found on many ranches. A glassy green wood, highly agatized or opalized, is a rare prize found near Valentine.

The fossil ivory of the region is incomparable. Hard, finegrained, pastel in color, and often subtly patterned with dendrites, mastodon ivory has been found in lakes and rivers, but is now very difficult to come by. Novices and sometimes even advanced collectors mistake the sections of ivory for petrified wood. The ivory is white, cream, pale blue, or even pale rose, due to mineral inclusions, but the wood here is almost always dark. If the wood shows growth rings, they are not as uniformly round and concentric as the structural rings of the ivory.

The South Platte and North Platte Rivers join at the city of North Platte in Lincoln County. Gravel deposits in all directions from North Platte have collectable quartz gems. The area between the two rivers west of North Platte has yielded a remarkable variety of lapidary material, as have the gravels east of

Xs show agates, jasper and petrified wood along the North and South Platte Rivers

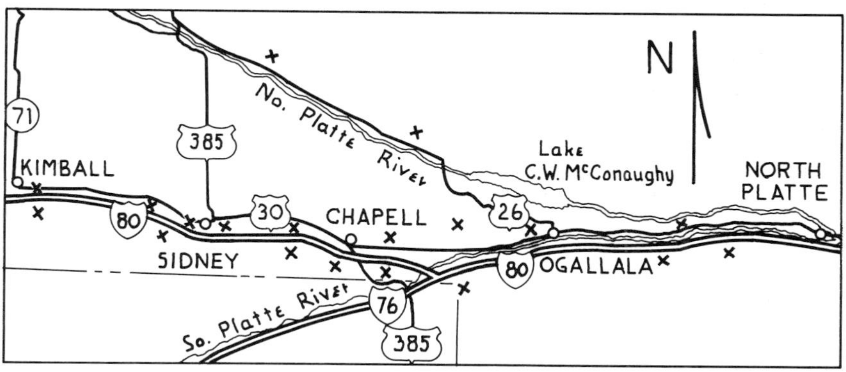

the city. The access route is U.S. 50. Interstate 80 also follows the river and there are 5 exits in the North Platte area.

There are many exposures for Pleistocene fossils south of the Platte River. Fossils are also found north of the Platte in the Sand Hills.

Railroad and highway cuts in Pleistocene formations of Gosper County have also yielded numerous fossils, both vertebrate and invertebrate.

Petrified wood is found south of Thedford in Thomas County. Colorful chert, locally called jasper, is found in Furnas County along the Republican River. Other quartz gems are present.

Eastern Nebraska

About 30 of Nebraska's counties were glaciated in Pre-Wisconsin times. These counties are all in the eastern part of the state in a three to four county-wide band parallel to the Missouri River. The lapidary's prime goal in the glaciated region is the colorful banded agate known as the Lake Superior agate. With the Lake Superior agates in the glacial tills are pieces of facetable quartz, many colors of jasper, moss agate, carnelian, and such non-quartz materials as feldspar and unakite. There are also some mixtures of quartz and hematite or other iron mineral, which resemble the chatoyant binghamite of Minnesota. Some of the petrified wood is opalized.

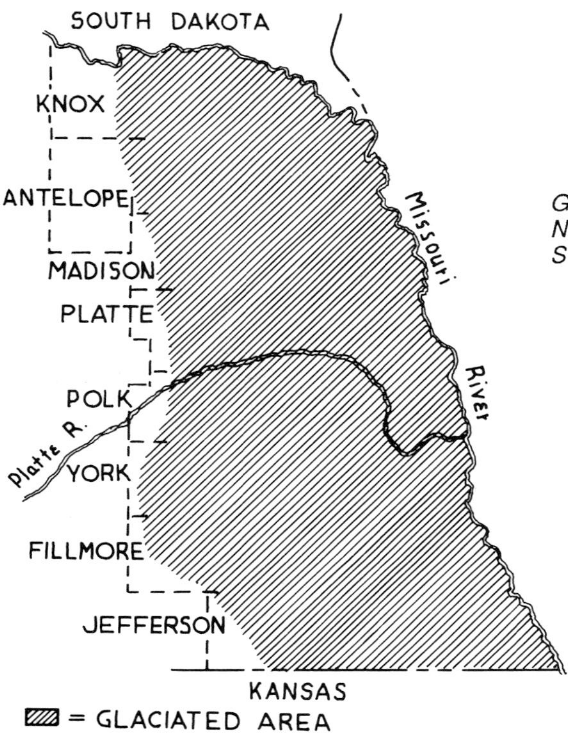

Glaciated regions of eastern Nebraska where Lake Superior agates are found

This Pennsylvanian trilobite, Ditomopyge, was collected near Weeping Water.

There are many rivers in eastern Nebraska, all with the possibility of collectable gems or fossils. These are the rivers: Niobrara, Elkhorn, Cedar, Calamus, North Loup, Middle Loup, Dismal, South Loup and Loup, Platte, Republican, Little Blue, Big Blue, Big Nemaha and Little Nemaha.

Selenite crystals occur southwest of Niobrara. Another gypsum location is in the Pierre Shale formation in Knox County south of the Gavins Point Dam.

The drainage area of the Loup River, which covers a large part of Central Nebraska is rich in gem materials. Feldspar with chatoyant luster, dark green epidote, and a wealth of quartz materials are found in the sand and gravel bars. There are many species of well-silicified wood. Howard County is a good location for the wood. Nebraska Highway 2 follows the Middle Loup River from Mullen to Dunning County and private roads lead to the river banks, terraces and bars.

Gem materials, marine fossils and crystals of selenite, marcasite and pyrite are found in South central Nebraska. Cretaceous outcrops in Nuckolls County near Superior, on the Republican River, have quartz gems, and fossils. There are similar areas along the Republican at Guide Rock in Webster County and Franklin in Franklin County. Look also along the Little Blue River in Adams, Clay and Thayer Counties. Lake Superior agates, honey agates, petrified wood, moss agate and marine fossils are among the "keepers".

Fossils such as clams, oysters and fish parts are also found south of Franklin. Cross the river on route 10 and turn west 1 mile on a county road.

Honey agates are banded in tints of yellow, cream, white, clear, and brown,

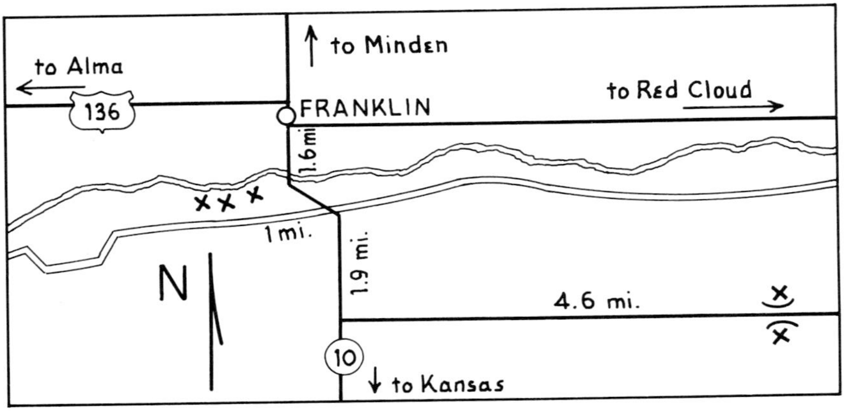

and sometimes have dendrites. They are opaque to translucent. The petrified wood of the area can be white, brown, red, orange or yellow.

The entire lengths of the Big Nemaha and Little Nemaha Rivers produce plant and animal fossils and lapidary materials. Lake Superior agates are found near Table Rock and Humboldt. There are snails, brachiopods, trilobites and nautiloids on the River terraces. These occur in limestone, while the sandstone has plant fossil imprints. Watch for the glacial deposits in road cuts and excavations, too.

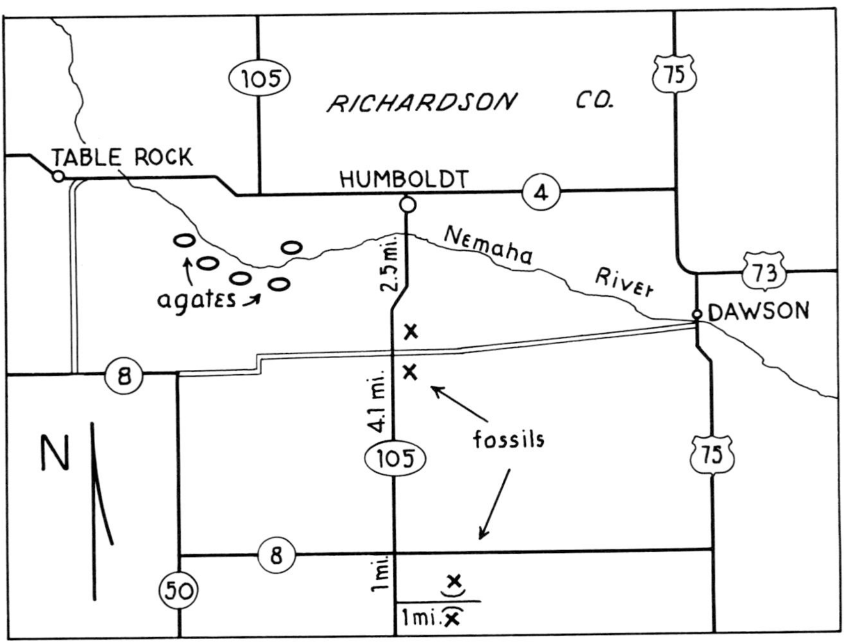

Roger Pabian, prominent geologist and writer, says that canoe trips on Nebraska Rivers are profitable to the collector. He mentioned several of these areas in his 1982 article on Nebraska gem materials for the *Lapidary Journal.* Gypsum and marcasite crystals occur 4.2 miles south of Fairbury. There are also plant fossils and invertebrate marine fossils here. One and a half miles east of Humboldt on Spring Creek there are Cretaceous fossils. Pelecypods are found south of the town of Nemaha.

Pabian advises prospectors to try to visit as many as 20 to 30 small outcrops in a day, as most of the gravel deposits are not massive. He also advises that the most productive sites will have a large proportion of dark-colored rocks. The tills to avoid are those with a high proportion of pinks, whites and reds.

A quarry just west of Weeping Water in Cass County has attractive horn corals and other invertebrates. Many fossils have been found in an old quarry 1 mile north and 1.7 miles east of Weeping Water. Another old quarry is 0.7 mile west and 0.1 mile south of the Highway 50 and S13K junction. Numerous fossils are south of Humboldt and along the Nemaha. Fossils such as bryzoans are found in a river bluff 1/2 mile north of Peru in Nemaha County.

Calcite crystals are found in limestone of the Pennsylvanian south of Humboldt. Calcite can also be found near Dunbar in Otoe County; and near Douglas and Burr in Otoe County, calcite geodes and gypsum occur in shale. Geodes from a quarry 3 miles south of Wymore are noted for crystals of blue celestite. Geodes are also exposed near Holmesville on the Blue River, southwest of town. Small but pretty geodes also come from Gage and Nemaha Counties. A notable geode location is 5 miles south of Blue Springs on the east bank of the river.

Near Odell in Gage County tiny elegant barite crystals occur, locally called "Odell Diamonds". The small crystals are rhomboid and tabular. Transparent to translucent and lustrous, they are pink, clear, and yellow in color. Phantom zones add interest to the crystals which are associated with the gray shale. The former collecting site is now closed, but crystals have been reported from other spots southwest of Odell.

Similar tabular barite crystals, but blue and gray in color, have been found in Cass County east and south of Weeping Water, near the Missouri River. Barite and calcite crystals have been found in a quarry south of Beatrice in Gage County.

Among the fine fossils of eastern Nebraska are trilobites, brachiopods, pelecypods, corals, nautiloids, bryzoans, crinoids and fish. Occasionally major finds excite the paleontologist such as the plesiosaur from Saunders County now in the University of Nebraska Museum in Lincoln, or the recent discovery of a herd of ancestral rhinoceroses at a watering hole which existed in Antelope County some 10 million years ago.

Hints for Collecting Invertebrate Fossils

1. Include the following in your equipment: prospector's pick, cold chisel, screwdriver, knife, paintbrush, spade, pry bar, sledge, acrylic spray, wrapping paper, notebook, pen, small bags and boxes, knapsack, hand lens, aluminum foil, steel tape.
2. In addition to the sites mentioned, other places to look for fossils are: peat bogs, ballast dumps, city dumps, canal dredgings, pipelines, cable trenches, ghost towns, spillways, mine washing plants.
3. If working in a shale pit, check the face of the stone layer by layer.
4. If the fossil is crumbly, spray it with dull finish acrylic or brush it with white glue diluted with water.
5. If the fossil is broken into several pieces, wrap the pieces together in aluminum foil.
6. If you find a fossiliferous layer, follow it along the outcrop.
7. Besides noting the site where you found the fossil, measure how deep it was, note in which part of the quarry, what other fossils were with it and other pertinent facts.
8. Collecting fossils is not an Olympic event. Work slowly.
9. Try to remove as much unnecessary rock as possible at the site.
10. Move atonce if pebbles start falling overhead.
11. Alwaysorient yourself so that you can find your way back to your vehicle. However, every field collector should study survival tactics.
12. In late summer, when waters of rivers or reservoirs may be low, many fossils have been found by collectors wearing bathing suits and rubber shoes, wading near the shore or from gravel bar to gravel bar.
13. When you get permission, always take time to visit with the land owner. Many times an owner will know exactly where fossils are, but has no interest in them.

Pearls have been found in the past in Nebraska rivers and spring-fed lakes. Pearl-bearing mussels are present in Carter Lake, the Big Blue River, and some lakes in the Sand Hills.

Nebraska has over 8O federal and state camping grounds with facilities for motor homes, trailers, pickup campers, and tents. The best months are in spring and fall. It cannot be emphasized too strongly that collecting cannot be done without permission and the AFMS Code of Ethics must be followed.

The State of Nebraska Geological Survey is located at the University in Lincoln. The Survey has an unusual number of excellent and useful publications which are up to date and full of information for those interested in any of the earth sciences. Many of these are by Roger Pabian who is an enthusiastic rockhound and collector, in addition to being a highly trained professional. There is also an excellent museum at the University of Nebraska, featuring many fossils which amateurs have helped collect.

There are also museums with earth science exhibits at Hastings, Kearney, Lewellen, Oshkosh, Gering, Chadron, Harrison, and Crawford. A noted artifact collection is in Weeping Water.

Nebraska has clubs divided between the Rocky Mountain and Midwest

Federations, and united by the Nebraska State Association. Several clubs have well-attended annual shows. The Crawford show features Nebraska materials. There are also rock swaps, picnics, club field trips and other activities.

Rock shows are scattered over the state in Omaha, Lincoln, Norfolk, Monroe, Chadron, Crawford, Valentine, Kearney, and several other cities.

Visitors are welcome at the Lincoln Gem and Mineral Society which meets at the University. This club is one of the few to twice host a National Show.

This Pennsylvania Age crinoid, Delocrinius was found in the Irvine Creek limestone near Union. Photo courtesy of the University of Nebraska, College of Agriculture.

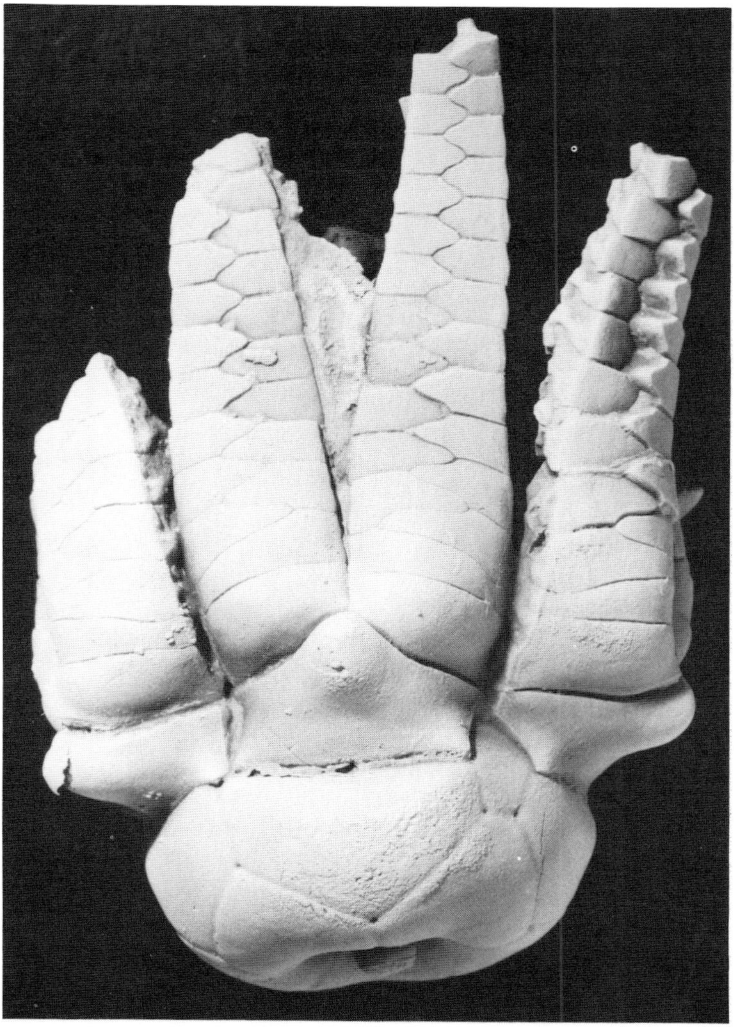

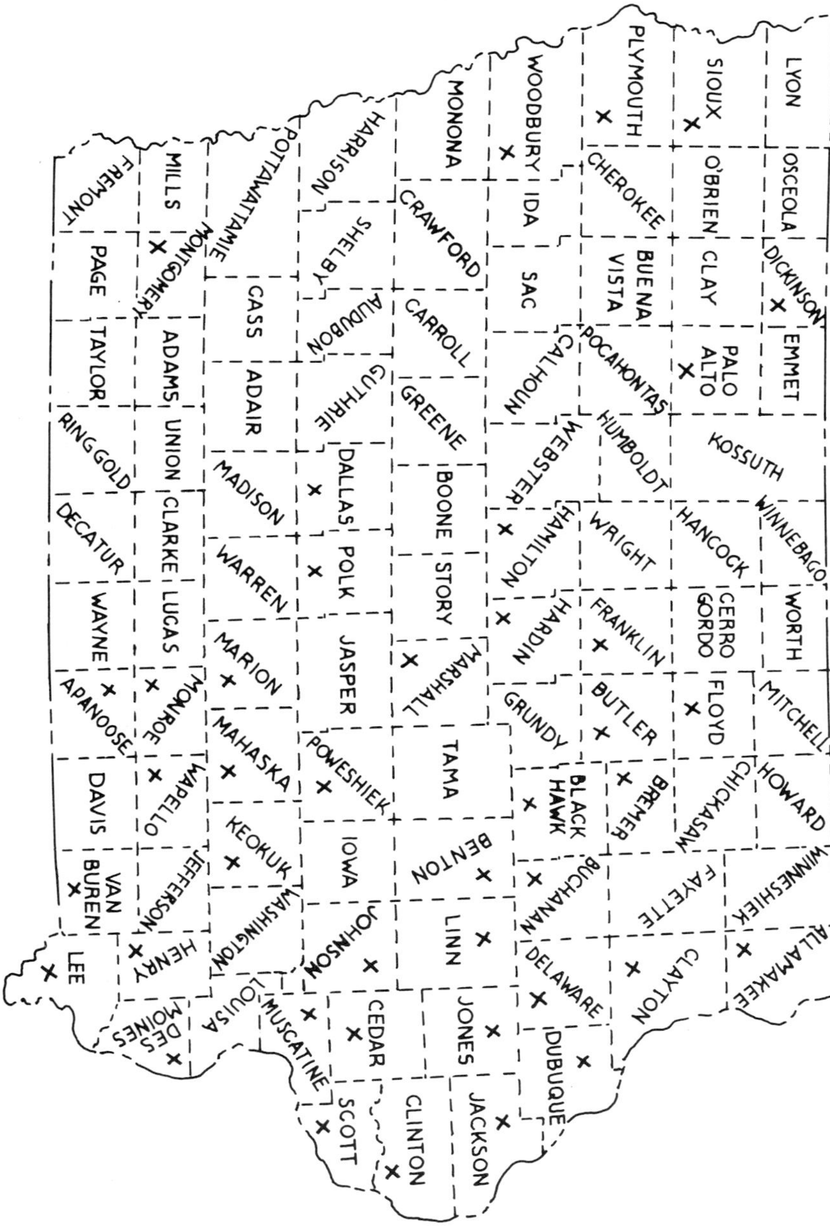

Xs indicate counties with gem, mineral and fossil locations

IOWA

Iowa, a fabled land of geodes and agate, is dotted with coal mines, quarries and gravel pits of great interest to the rock hobbyist. Iowa is a prairie state bordered on the west by the Missouri River and on the east by the Mississippi. Both borders are marked by steep bluffs and rolling hills. Evidence of glaciers is widespread. There are numerous well-located State Parks, but the rest of the land is private, so permission to collect is necessary.

Northern and Central Iowa

Northwestern Iowa has agates and silicified wood, but they are widely scattered with no concentrated areas. Sioux County has reported agates of the Lake Superior type and agates and jasper are found along the Little Sioux River. Fossil fish scales and bones, and pelecypods are found in quarries near Sioux City.

Agatized wood is found near Emmetsburg in Palo Alto County; Lake Superior agates are found around Five Island Lake and along the West Des Moines River.

Interesting concretions cemented by dark hematite are found along the Big Sioux River in Plymouth and Woodbury Counties. Also along the Big Sioux it is possible to find pieces of pink to red Sioux quartzite, and sometimes pieces of carvable dark red catlinite, the pipestone of the Sioux Indians.

In Hamilton County near Webster City there is an excellent outcrop of cone-in-cone calcite. The well-formed nested cones are deeply striated and jet black in color. The location is near Bells Mill Park, 5 miles north of Stratford, where County Road "E" crosses the Boone River. The deposit is on the west side of the river. Agates are also found in the glacial gravels of this county.

Lake Superior agates, carnelian, chalcedony, feldspar and other lapidary materials are found in the gravels of the rivers and streams and quarries of Polk County. Colorful banded agates have been found in the city limits of Des Moines. Look along the Des Moines, Skunk, and Raccoon Rivers and around Big Creek Lake.

Dallas County has some striking gypsum crystals, some with black carbonaceous inclusions. The location is along a county road 3 miles north of Dexter near the South Raccoon River. Another location is north of Granger on the west side of the Des Moines River valley. Pyrite and marcasite are found in the coal seams of Dallas County; showy clusters of bright pyrite crystals are found in blue clay in stream banks south of Adel.

North central Iowa locations in Franklin, Butler, Bremer and Hardin Counties have geodes. Quartz geodes are found 1 mile east of Chapin on Highway 65 and south of Baileys Creek at Sheffield. In Hamilton County geodes are found 2.5 miles north of Roland. The Butler County location is south of Hubbard 5 miles and then west and southwest on county roads.

Look for gypsum crystals in Franklin County southeast of Sheffield 4 miles, east of Beaverdam Creek. An old quarry at Hampton on Spring Creek has numerous fossils, while quarries near Sheffield have polishable silicified

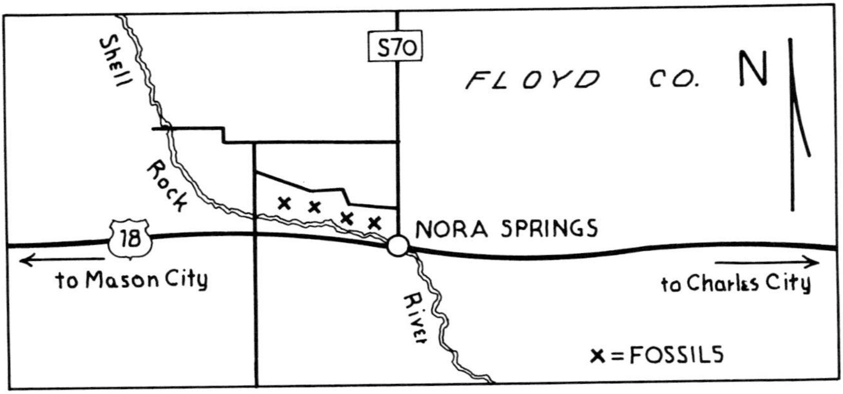

Fossils along the Shell Rock River northwest of Nora Springs

fossils. Fossils are also abundant at a quarry near Chapin where crinoids, corals and brachiopods are interbedded with chert.

In Hardin County look along the banks of the Iowa River near Eagle City for brachiopods, bryzoans, and corals.

A location for marine invertebrate fossils is in Floyd County along Shell Rock River northwest of Nora Springs. Brachiopods, corals and crinoids are found here near the park.

Black Hawk County in northeastern Iowa has long been a mineral collector's dream. The locality for esthetic crystals and lovely agates was closed for several years, but it is reported that groups with insurance, safety equipment and guides are admitted for collecting. The site, known as Pint's Quarry, is east of Raymond, a small community 5 miles east of Waterloo on U.S. Highway 20. Dogtooth calcite, crested barite, amber to brown fluorite, pyrite, galena, sphalerite, and drusy quartz are some of the glamorous crystals from this long popular quarry.

Coldwater agate is found in Black Hawk, Benton, and Bremer Counties. A lacy type of pastel fortification agate, coldwater agate occurs in nodules of chert. The colors are white, ivory, beige, ecru, sepia, gray, yellow, orange and

Quarries near Waterloo have crystals, fossils and cutting materials

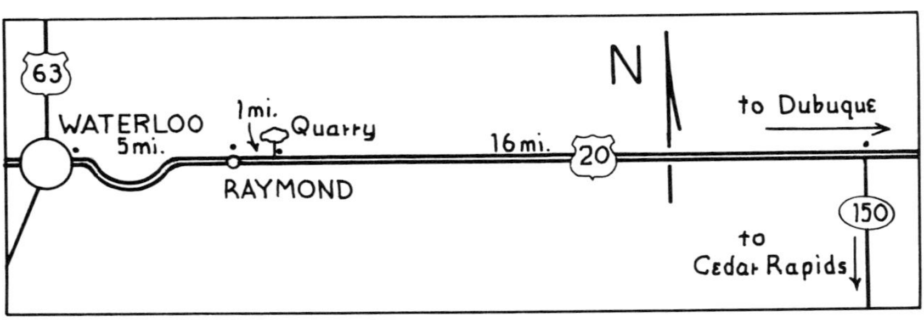

pink. Sparkling geode cavities or botryoidal linings often occur in these agates. The agate bands may be alternated with crystalline quartz. Iowa lapidaries have made stunning cabochons and jewelry of this native material which has been found along the bluffs of the Cedar River in northeastern Benton County. A productive location is 2 miles south of Urbana in Benton County while another location in this County is near Shellsburg.

In Linn County coldwater agate is found near Walker. This type of agate, with some variations, has been found in numerous quarries road cuts, stream banks and excavations . The agate-bearing formation often starts about 2 feet beneath the surface. Silicified fossils are also exposed in Linn County.

Intricate and exquisitely preserved colony corals are found in Buchanan County. The fossil corals are also found in the valleys of the Iowa and Cedar Rivers in Black Hawk, Benton, Cedar and Johnson Counties. Many of the corals are Hexagonaria. Some of the corals are silicified and have attractive pastel tints, making them desirable cutting materials. Some of the locations are east of Independence, north of Vinton, and near Cou Falls. Probably the best lapidary quality material has come from a closed locality near Coralville in Johnson County.

Fascinating agate geodes are found in stream beds and banks near Manchester and Delhi in Delaware County. Apparently pseudomorphs after marine life, the chalcedony and quartz geodes are banded similar to the coldwater agates. Graptolites and trilobites are found 2.5 miles west of Manchester.

A rare lilac-colored phantom calcite comes from Poweshiek County. The unusual calcite is enhanced by inclusions of bright, finely disseminated pyrite. The original location is 3.5 miles north and 1 mile west of New Sharon.

A world-famous fossil quarry is in Marshall County at Le Grand. Here Dr. B.H. Beane, an amateur paleontologist discovered, collected and prepared many remarkable crinoids, starfish and other museum quality specimens. Some of these fossils are now on display in the Iowa State Historical Museum in Des Moines. Exposures along the Iowa River in this County are very fossiliferous. Look in the blue siltstone ledges for gastropods, brachiopods, and pelecypods.

Allamakee, Clayton and Dubuque Counties in northeast Iowa all have old lead and zinc mines. Galena was found as float in alluvial beds of stream valleys in the early years and lead ore occurring in vein deposits was mined in the mid-1800s. Attempts were made as late as 1955 to reopen some of the old mines.

The time to look for the minerals is in the spring before the ground cover gets too dense, and preferably after a hard rain. This is a rugged and scenic area of forested bluffs and steep river banks.

Small pieces of galena and other minerals can still be found in the dumps of old mine workings along South Catfish Creek and in the Durango area. The old Beadle Mine dump is located on the south side of Palmer's Creek. In this area, galena is also found in some of the sand and gravel quarries. Some of the mines are near North Buena Vista and Guttenberg in Clayton County while the Holmes Mine is 2.5 miles up Miners Creek Road from the south edge of Guttenberg. Other mines are north of Waukon and northwest of Lansing in Allamakee County. The greatest activity here was on Mineral Creek 8 miles

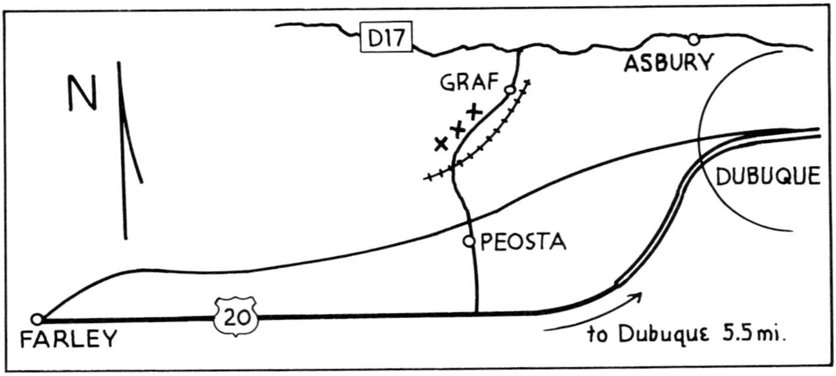

Gastropods and cephalopods in shale bank southwest of Graf

due north of Waukon off Highway 76 and in the area 5 miles north of Lansing.

These areas yield: galena, sphalerite, smithsonite, pyrite, marcasite, chalcopyrite, azurite, malachite, anglesite, cerussite, calcite, dolomite, barite and gypsum. Collectors have also found limonite, chert, and drusy quartz. Permission is required at all of these sites and collectors should not venture into old tunnels.

The banded type of calcite commonly called Mexican Onyx is found in Dubuque and Clayton Counties. Violet and yellow fluorite cubes were found near Volga in Clayton County. Trilobites are found 1 mile southwest of Elkader in the Turkey Creek exposure there. Numerous marine invertebrate fossils are found near Graf and Florenceville in Dubuque County. The Graf location is on the west side of the road southwest of town and north of the railroad crossing.

Limonite is found at Iron Hills, northwest of Maquoketa in Jackson County and in the glacial drift southeast of Maquoketa. Look along Brush Creek, Farmers Creek and Lytles Creek, all north of the Maquoketa River as well as around Iron Hills. Pieces of galena and associated minerals have been found in the glacial drift of Jackson and Jones Counties.

Glistening selenite crystals occur on the west side of Highway 52, 1.5 miles southeast of Bellevue State Park in Jackson County. Graptolites are found along Mill Creek north of the Park.

An interesting quarry for fossils is northeast of Maquoketa. Take Highway 61 to 1 mile north of its intersection with Highway 64 and turn east just south of Hurstville. This road leads to the quarry about one-half mile from the highway. Brachiopods with unusual details are found here. Get permission.

A type locality for pyrolusite dendrites in the United States is in northern Scott County near Princeton. The graceful dendrites are found in quarries in the Niagaran dolomites of the Silurian age. The dolomite is a light yellowish brown and the decorative patterns are "inked" in a warm sable brown shade. One quarry is one-half mile east of Highway 61 and 12 miles north of Bettendorf in Scott County. Another is 1.5 miles south and 2 miles west of Grand Mound in southern Clinton County.

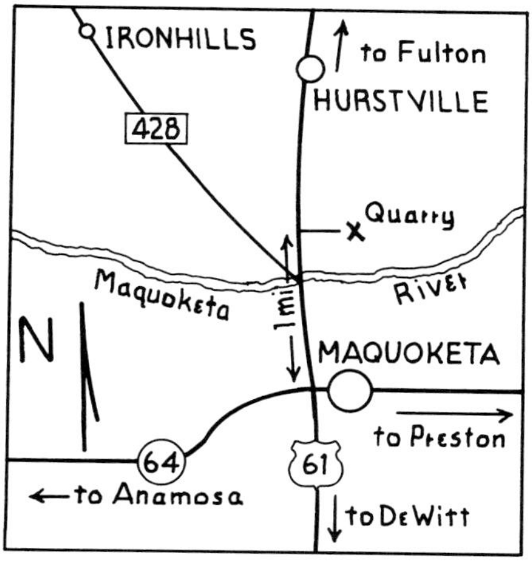

Marine fossils in quarry north of Maquoketa

Elegant slabs are often used by decorators as focal ornaments in modern buildings.

Trilobites and other marine invertebrates are found in quarries near Davenport. Graptolite and trilobites occur in the limestone near Clinton.

Dendritic limestone near Clinton

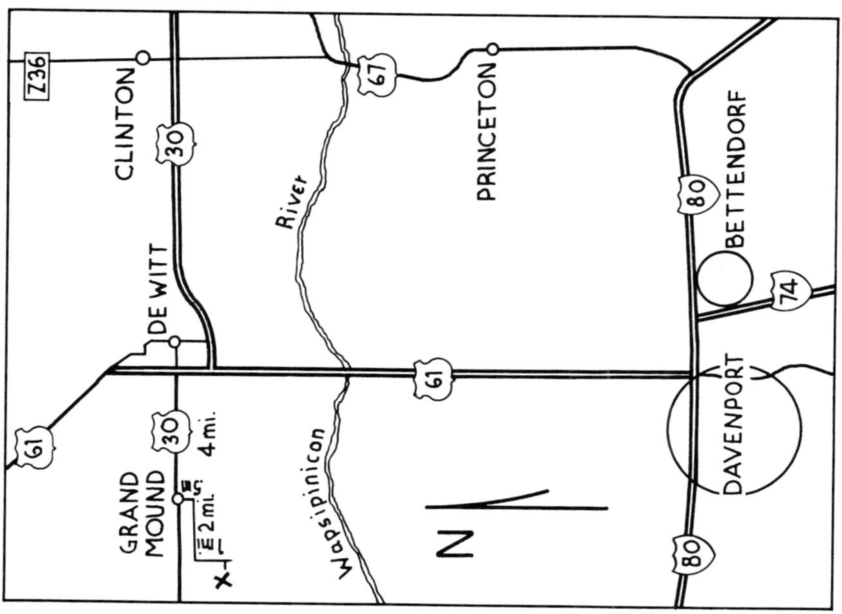

Southern Iowa

Montgomery County has an interesting and polishable cherty limestone, locally called "Rice Agate", or sometimes "Protozoa Agate". The chert is dark brown, gray, charcoal, or black with white or cream-colored patterns of small white fusulinid fossils, mainly triticites. The small fossils are variously shaped, but the oval or rice-shaped ones predominate. The chert is present in quarries near Red Oak, Stennet, and Grant. Well-polished selected pieces make striking cabs.

Puddingstone conglomerate, locally called peanut conglomerate is also found in Montgomery County at Coburg and along the edge of the Nishnabotna River valley. Small pebbles of various rounded pebbles are cemented together by the mineral limonite. It is found in areas of Cretaceous sediments and can also be seen in Cass County, north of Montgomery County along a tributary to Seven Mile Creek.

An old gypsum mine is at Centerville in Appanoose County. With proper permission to search, gypsum and prismatic celestite crystals can be found in dumps. Clay suitable for pottery is another product of Appanoose County and pyrite and marcasite are found in old mines. This is an old coal mining region and there are countless dumps in the county from Moravia on the north border to Coal City on the south where some plant fossils may also be found.

Several counties in South central Iowa have produced spectacular crystal specimens from large concretions found in the strip coal mines. These adjacent counties are Marion, Monroe, Mahaska and Keokuk. Some of the most beautiful specimens are dolomite, white, yellow, brown, and iridescent dark calcite, white barite, crested barite, and sometimes pyrite, marcasite, and goethite. The showiest specimens have sharp pineapple-shaped clusters of white calcite perched on iridescent peacock-colored drusy crystals coating the limestone of the concretion sections.

Puddingstone conglomerate location near Red Oak, Iowa

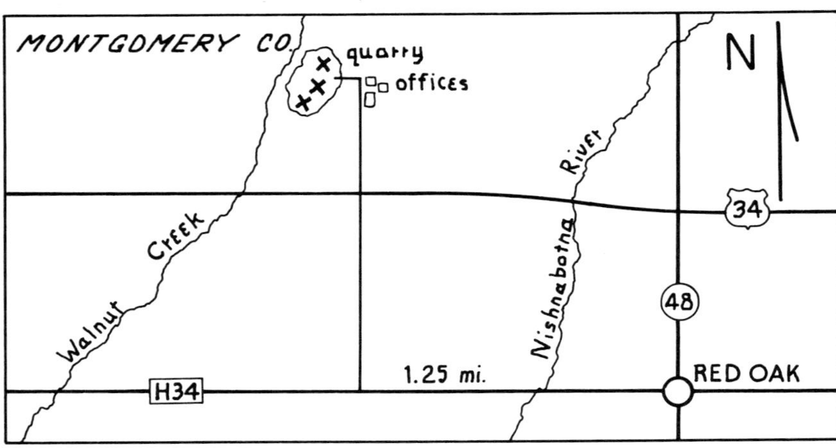

Pink calcite is one of the many inclusions to be found in the geodes of Iowa and Missouri.

These graceful pyrolusite dentrites adorn the dolomite in several eastern Iowa quarries.

Finely chiseled chert arrowheads are found in most Iowa counties.

Crinoid fossils from Marshall County, Iowa are among the finest known.

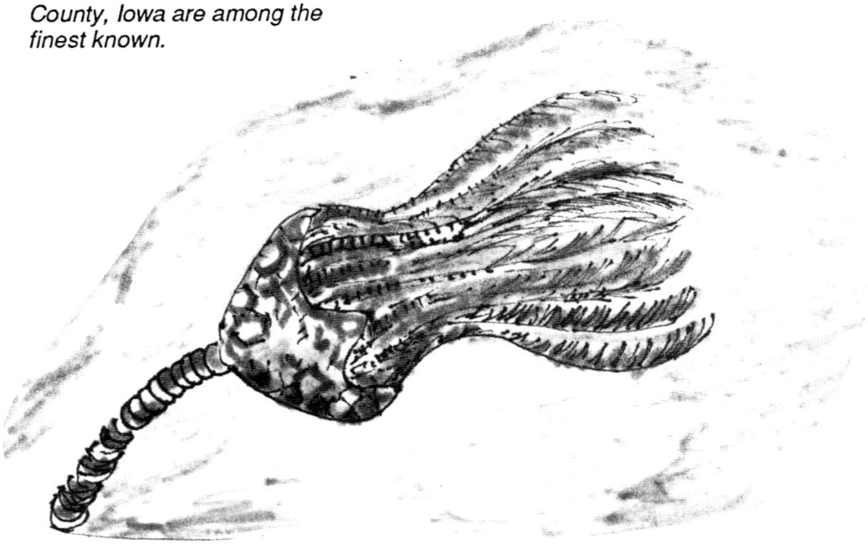

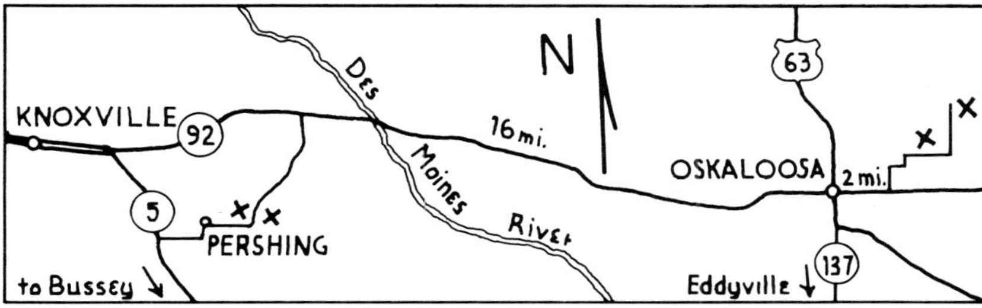

Locations for crystals and fossils near Oskaloosa

The "coalball" concretions are found in many parts of this region. However, several areas are now closed. The Pershing Mine between Knoxville and Oskaloosa was a prime locality for exotic specimens. The mine is west of the Des Moines River and south of State Highway 92, 16 miles west of Oskaloosa. From Knoxville, the mine road is the first trail going south after passing the Red Rock Dam road which goes north to Pella. There are many mines in the area. Local inquiry should be made and permission to collect is needed. Another old mining area is northeast of Oskaloosa, beginning 2 miles east on the north side of Highway 92.

In these same mines rare fossil wood is found; most highly prized are the huge sections of black lepidodendron, intricately patterned with amazing details preserved in calcite, iron, and pyrite, a stunning material when polished.

Other mines are near Tracy, 2.5 miles east of Oskaloosa, 5.5 miles east of Bussey, and 6.5 miles west of Eddyville. The strip mines continue into Wapello County. Brachiopods, corals, crinoids and other fossils are found in quarries near Ottumwa.

A rich site for fossils is north of the Skunk River in a quarry east of Highway 63 and just south of a road joining route 63, 5 miles north of Oskaloosa. Turn east and look for the quarry just after you cross the railroad tracks. Trilobites, crinoids and bryzoans are some of the fossils in this location.

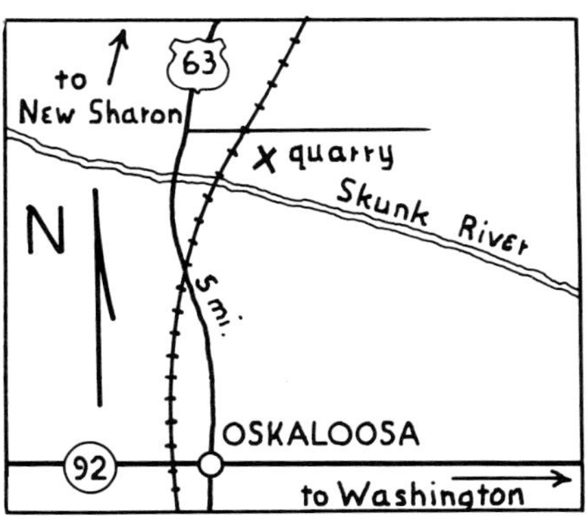

Marine fossils in quarry north of Oskaloosa

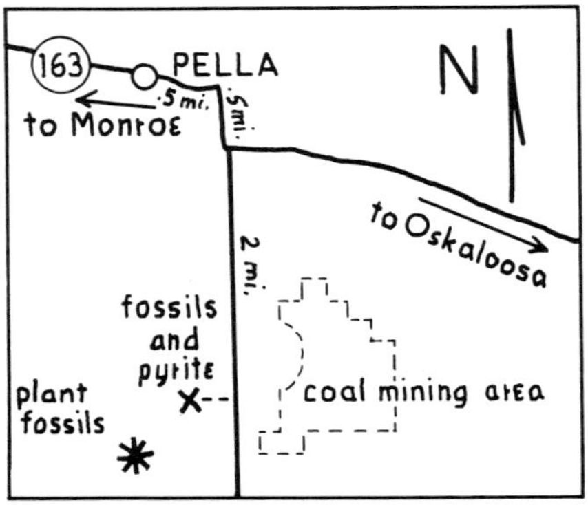

Fossils and pyrite south of Pella

Another location for fossils is southeast of Pella. Take Highway 163 east and south of Pella for 1 mile. Turn south here. The quarry is on the west side of the road 2 miles south of Highway 163. On the dumps here are blastoids, bryzoans, brachiopods and other marine invertebrates. There is also some pyrite here. A little farther southeast there are Pennsylvanian plant fossils in an old pit now used as a dump. On the opposite side of the country road there is an old coal mining area.

Many of the old quarries are water-filled and the fossils or minerals may be on steep cliffs near the water. Collecting in such places is extremely dangerous. Collectors have found two solutions. If they have permission and can get a good boat, they can row over to the fossil horizon and collect pieces of rock to take home and then work the fossils out there. The second solution is to wait until winter when there are 7 or 8 inches of ice or more on top of the water. Then, dressed for the weather, collectors walk around the edge of the ice and dig fossils from the quarry walls. The ice must be thick! Collectors should not enter such areas without a companion.

In Keokuk County, the Ollie quarry has produced Iowa's finest millerite specimens. The acicular shining crystals of dark millerite, sit at jackstraw angles on white or clear quartz. Pyrite sometimes brightens the specimens. The crystals occur in vugs in limestone at Ollie and in several other Iowa quarries. Millerite is also sometimes found in geodes. Millerite has also been found west of Farmington in Van Buren County and North of New Sharon in Mahaska County.

A notable gem of Keokuk County is Keswick agate, a pastel fortification agate which has brighter colors and often wider banding than coldwater agate. Bands of bright peach, carnelian, hot pink, or tomato red parallel the paler bands of white, ivory, sand, and cream. Drusy quartz crystals sparkle on many of the interiors. The Keswick agate is in cavities in limestone and chert. The quarries are near Keswick, Ollie, and Harper. Trilobites are found near Ollie and Fremont.

Muscatine in Muscatine County was for years one of the best spots in the nation for bright and wonderful Lake Superior agates. Some of the most productive gravel operations which produced these beauties are now closed, however agates are still found in the Muscatine area. Lake Superior agates

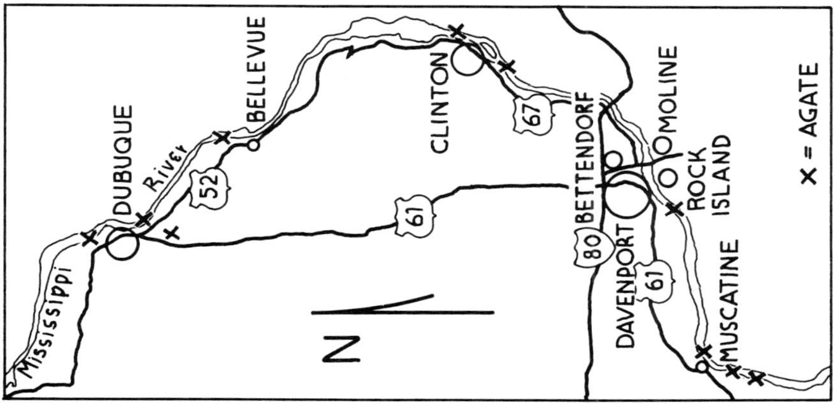

Agate locations along the Mississippi Rover

and a host of other fine lapidary materials are associated with any gravel operation along the Mississippi River in Iowa. But the agates are not only in the gravel piles of commercial operations. There are agates to be found along roads, streams, ditches, plowed fields, railroad tracks, and in excavations throughout the county...virtually anyplace where permission to collect is available. Besides "lakers", other materials for lapidaries along the Mississippi are moss, dendritic or "Montana"-type agate, white, cream, and gray chalcedony, chatoyant pink, peach and ivory feldspar, silicified wood of various species, corals, carnelian, onyx-banded agates, epidote, crinoidal limestone, conglomerate, breccia, and pieces of white, smoky, or yellow quartz. A hard dolomitic shale is locally called "fishtooth conglomerate". Polishable hexagonaria coral is found at a quarry in northern Muscatine County south of the Cedar River at U.S. 6 near the town of Atalissa. (Hexagonaria is the much

Atalissa Quarry for polishable hexagonaria 1 1/2 miles south of Exit 265 from Route 80

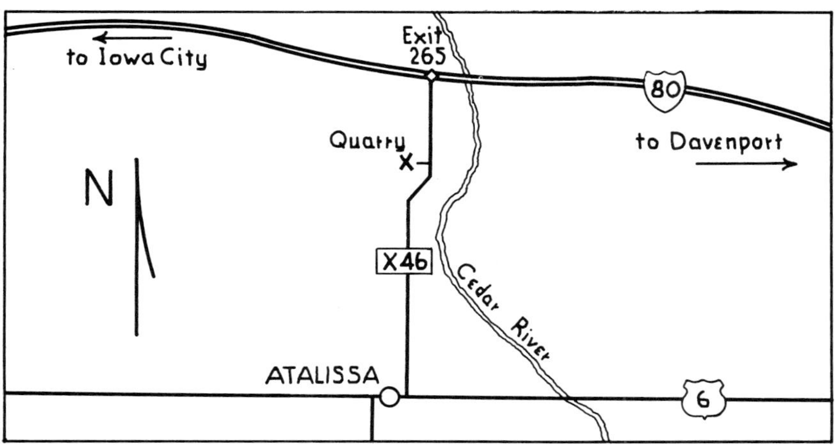

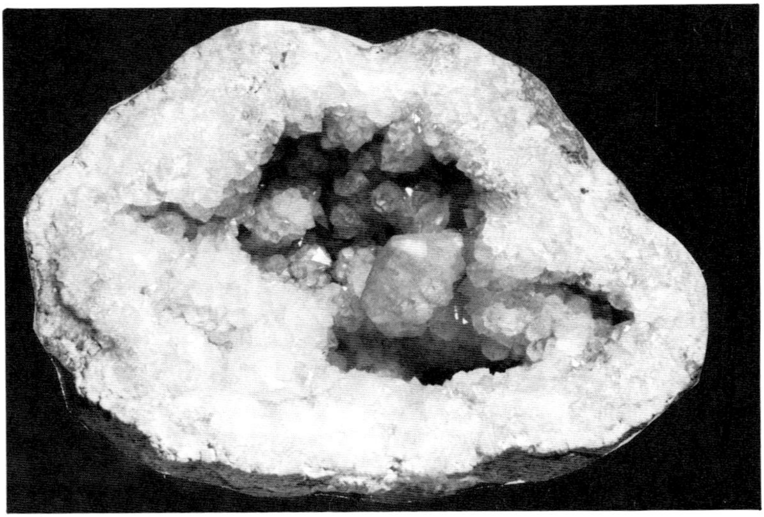

Quartz geodes from Warsaw formation are found in many Iowa locations

admired Petoskey stone of Michigan.) With the corals are many species of marine invertebrates including forams, sponges, stromatoporoids, bryzoans and echinoderms.

The town of Muscatine, a haven of agates, became famous as a boom town for gems more valuable than agates many years ago. It was the center of a great "pearl rush" and the resulting pearl industry of the prairie states of the Midwest.

The lovely freshwater pearls of the Mississippi River and the thick iridescent "mother-of-pearl" aragonite became the basis of major industries here and in other river cities. The extremely lustrous and delicately tinted pearls commanded steep prices and were fashion's favorites. Factories in Muscatine manufactured pearl buttons, cuff links, shirt studs, buckles, hair ornaments, crosses, brooches, boxes, inlaid objects, and countless other items from the high quality mother-of-pearl. Local collectors have examples of some of these items which are now classed as antiques.

The pearls were found in a family of mollusks called Unionidae. The species living in the Mississippi River prefer swift flowing water. They live on the bottom in areas where the depth of the water is fairly constant.

One method of taking the mollusks was to drag the bottom with barbless hooks on lines suspended from a long pole. Elongated tongs were also used to reach for the shellfish. Many of the shells had baroque pearls. Mollusks still live in the Mississippi and lovely pearls are still found.

"Keokuk geodes" of the Warsaw formation of southeast Iowa are well-known among collectors of minerals in the United States, and these glamorous hollow rocks deserve their fame. They are truly nature's surprise packages, sources of wonder and pleasure for the lovers of crystals and esthetic perfection. Estimates of the number of mineral inclusions in the Iowa geodes vary from 16 to 40. Most geodes are quartz, chalcedony and calcite. Other inclusions may be pyrite, marcasite, sphalerite, galena, hematite, limonite, fluorite, dolomite, goethite, kaolin, opal, selenite, ankerite or millerite. Glittering metallic pyrite, pink calcite, clear quartz dew drops, and slender hairs of millerite are among the most glamorous.

Geodes vary greatly in size. Small ones may be about the size of a walnut, while large ones can be compared to big pumpkins. However, average sizes are somewhere between oranges and grapefruits. In the country where geodes are prevalent, they are used for walls, fountains, rock gardens, birdbaths, planters and door stops.

Many farmers will allow collectors to hunt for a modest fee, although there are no advertised fee areas at present. Butch Lewis of Bonaparte, Iowa (R.R. 1, 52620) has a guide service for geodes.

Geodes weather out of the limestone and are found in streams and rivers, road cuts, quarries and excavations in large numbers. Warsaw formation outcrops with geodes and fossils are found over a wide area. There are often fee basis collecting sites in this region. They are very productive if bulldozers are used periodically. Look for geodes along the Des Moines and Skunk Rivers and their tributaries, and in quarries in Lee, Van Buren, and Des Moines Counties.

Xs indicate geodes with calcite inclusions found in the area where Iowa, Illinois and Missouri meet

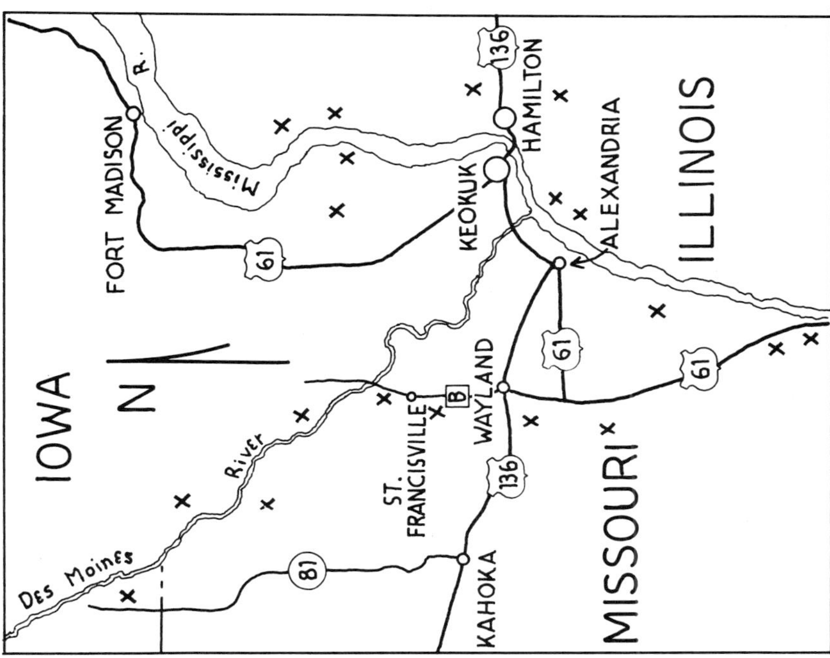

Some of the best-known sites are the banks and stream beds of Mud Creek 1 mile northeast of Lowell in Henry County. The Des Moines River near Farmington, Copperas Creek 2 miles northwest of Bentonsport and Indian Creek west of Farmington.

Geodes cannot be legally collected at Geode State Park south of New London but there are good geode exposures east and west of the park. Geode-studded limestone forms precipitous and dangerous river banks in some places.

With practice, a collector can learn to distinguish the "probably hollows" from the "probably solids". Then it is best to take them home before breaking or sawing them. Geodes are the State Rock of Iowa.

Agates, silicified corals, wood, jasper, chalcedony and other lapidary materials are found in the same areas as the geodes. A banded gray, charcoal, and red chert is found near Mount Pleasant in Henry County and in Big Creek and Cedar Creek.

Near Sperry in Des Moines County chatoyant gypsum known as "satin spar" is found and fossils are abundant near Burlington. With the plentiful crinoid stems, there are brachiopods, gastropods, pelecypods, bryzoans and corals.

Some of the rare minerals of Iowa include 1 verified diamond and several reports of small diamonds being found in glacial gravels. The single known diamond was gem quality and an apple green color about the size of a pea. It was found in the gravels of Dubuque and was sold to a Chicago jeweler.

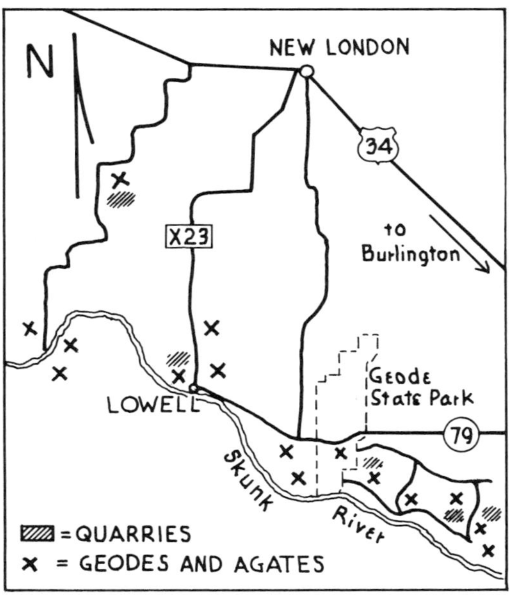

Geodes and agates south of New London

Crinoids (sea lilies) are among the most esthetic fossils, and those from Le Grand, Iowa are especially well known. Crinoids can be found in several Iowa quarries. Zeitner Collection.

Another rare gem for this part of the county is a corundum crystal, a variety of sapphire, which was found in 1912. After being cut it weighed 1.25 carats. Found at Lake Okoboji, Dickinson County, the gem was royal blue in color.

An unusual museum is the mineral and gem museum at Drake University in Des Moines. The Straight Collection there is maintained by the Central Iowa Mineral Society. The Iowa Historical Museum has fine Iowa specimens and is also in Des Moines. Other earth science museums are in Cedar Falls, Cedar Rapids, Dubuque, Iowa City, Mt. Vernon, Sioux City, Waterloo, and Waverly.

There are several active and influential clubs in Iowa, which draw enthusiastic members from all over the state. Big shows have been produced in Des Moines, Iowa City, Cedar Rapids, Davenport, Sioux City, and other places. The rock grotto at West Bend is on the "must see" list for Iowa visitors interested in gems and minerals.

Iowa rock shops are located in Cedar Rapids, Des Moines, Dubuque, Elkader, Keokuk, Sheldon and Stuart. There are also manufacturers and mail order companies in the state.

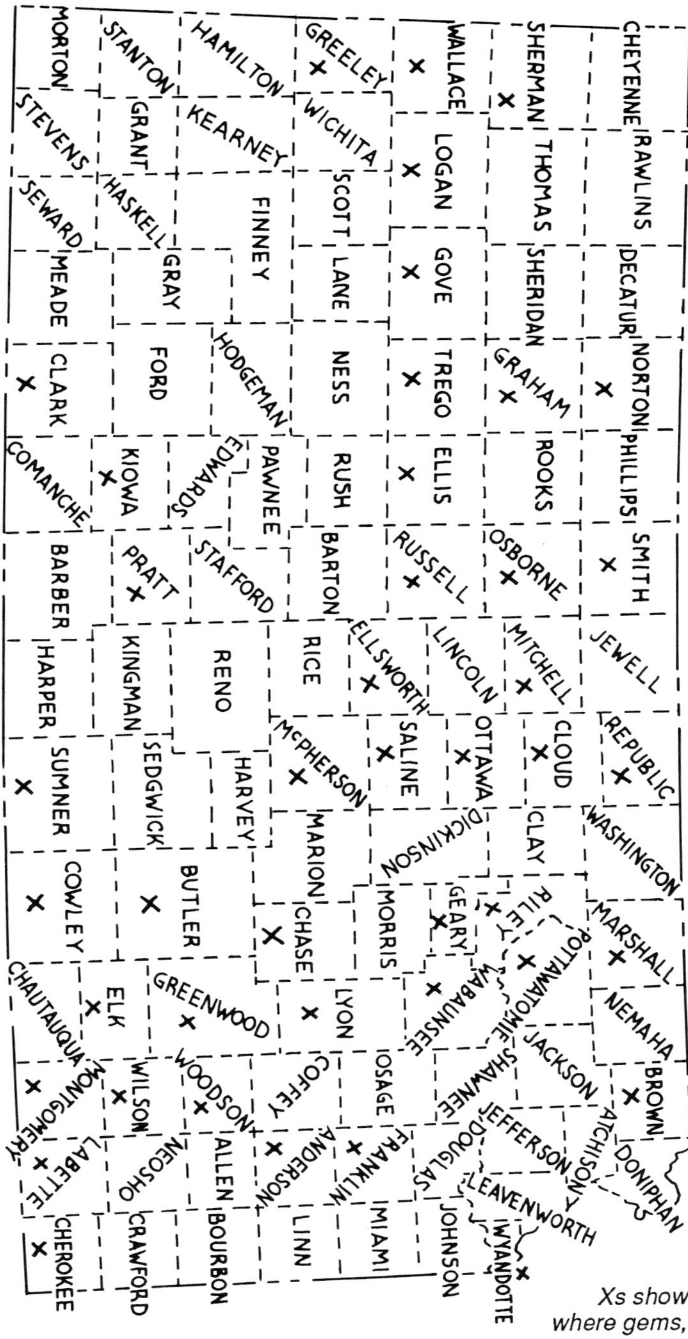

Xs show counties where gems, minerals and fossils are found

KANSAS

Kansas is a centrally located state combining features of both the east and the west. There is an assortment of collectable materials in the state, dominated by a large variety of fossils. The most famous mineral locality in the state, part of the Tri-State zinc-lead district in the southeast corner, is now closed to collecting. Gem materials, mostly quartzes, are widely scattered on private land. Kansas is a high ranking producer of oil, gas, sand, gravel, chalk, clay, limestone and diatomaceous earth.

Western Kansas

Western Kansas is a high and hilly country, sometimes with spectacular scenery of mesas, buttes, and spires. The Great High Plains are capped with Pliocene lavas, and there are many eroded Cretaceous exposures.

Common opal is a popular lapidary material of western Kansas. A good grade has been found 5 miles south of the town of Wallace in Wallace County. The opal nodules have a chalky white exterior and delicate black dendrites against white or colorless opal on the interior. Agates are also found in this county.

Wallace County has several fossil locations worth noting. On Goose Creek 1 mile from its junction with the Smoky Hill River there are numerous fossils in large limestone concretions. Vertebrate fossils have been found on

Moss opal, concretions and geodes near Wallace

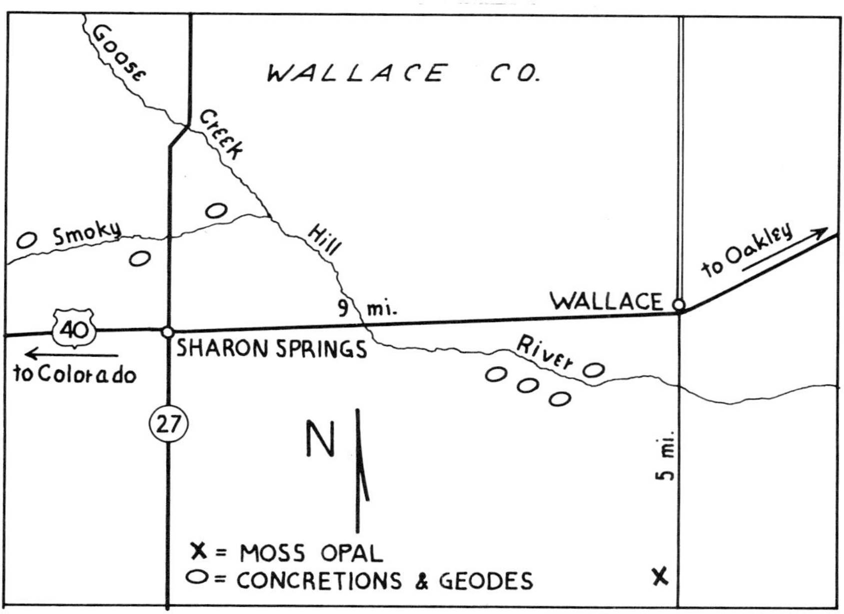

Rhinoceros Hill in the northeast area of the county. Near Sharon Springs, Eagle Tail Creek has marine fossils, also hidden in big concretions.

Agates, chalcedony and jasper are found along the Smoky Hill River. There is also a bright, patterned dendritic chert which is cuttable.

Common opal is found in Logan County near U.S. Highway 83. The opal is found in the chalk bluffs along the Smoky Hill River, a scenic and sometimes rugged area. Opal has been found near Elkader, an old community name in the county. Barite and calcite are found in some of the concretions near the river and its tributaries. Gypsum crystals and rosettes sparkle in a few spots.

Remarkable fossils have come from the chalk deposits of this county and its neighbors, among them fish, birds, plesiosaurs and mesosaurs. Fossiliferous concretions are found 2 miles east of McAllister Buttes. Marine fossils are found 3 miles south of Elkader and along the railroad 1 mile east of Lisbon.

The chalk formations of Gove County have vertebrate fossils of the Cretaceous Era along the Smoky Hill River and Plum Creek. Well-known areas are Monument Rocks and Castle Rock. Fossils are also found along Plum Creek and other tributaries to the Smoky Hill River. Lower Permian exposures along the waterways of Gove County have pelecypods and other marine fossils. These renowned fossiliferous formations continue into Trego County, where the Cedar Bluff reservoir affords collecting opportunities.

Gove County and Trego County both have opal, chalcedony, jasper,

Shark teeth, crinoids, shells and other fossils along the Smoky Hill River in west central Kansas

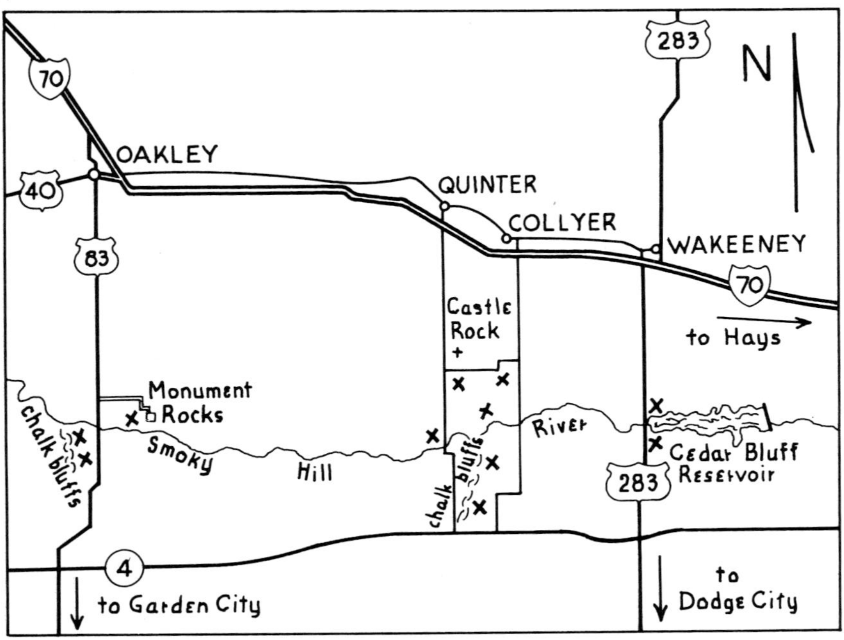

Septarian concretions have been found along the Smoky Hill River.

petrified wood and agates in the gravels. Calcite is found in some of the concretions; selenite and pyrite are other minerals of the area.

In Russell County fossils and agates are found along the Saline and Smoky Hills Rivers 5 miles northwest of Fairport on Wolf Creek, and around Wilson Lake. Banded travertine (aragonite) occurs on the Solomon River in Mitchell County.

In west and west central Kansas look for assorted quartz gems along the Arkansas, Cimarron, Pawnee, and Medicine Lodge Rivers. Stream banks, river bars and terraces, and alluvial gravels have a variety of interesting materials. Road cuts, quarries and gravel pits, and stock dams are other places to hunt...with permission, of course.

Concretions are found in Norton and Osborne Counties. Ellsworth County has an intriguing occurrence of cone-in-cone calcite. Resembling poorly petrified wood, the sedimentary structure occurs in the banks of the Smoky Hill River.

Celestite has been found in Ellsworth County, in blue or pink crystals or fibers, on the edge of the Kanopolis dam. Pyrite concretions come from north of the reservoir. Amber was found in the lignite beds which were exposed when this dam was being built. Barite and selenite have been reported in this county.

Moss agate is found north of Ashland in Clark County near Mt. Casino. Look along the banks of Bluff Creek and other tributaries to the Cimarron River.

The Post Rock country of west central Kansas is rich in history if not in gems. When the pioneers settled this part of the state in the nineteenth century, they found great, treeless plains. Industrious Germans, they wanted to build permanent structures, schools, churches, houses, barns, bridges and fences. So they quarried a native soft limestone, cut it into bricks, blocks, and posts and built everything they needed. The stone hardened in the air. Miles and miles of limestone fence posts still stand. These can be seen from highways and side roads from Hays to Russell, north to the Nebraska border and south to Great Bend.

Eastern Kansas

One of the best locations in the state for quartz family gem materials is north of Concordia in Cloud County. A variety of colorful cutting quality stones are found along the Republican River and in old waterways, and on gravel bars and terraces.

Barite roses are found in a quarry 2 miles east of Bavaria, a small community off State Highway 140 in Saline county. Chatoyant fibrous gypsum, commonly known as "satin spar" is found near the town of Gypsum on State Highway 4. Gypsum has been mined in this area. Satin spar is used for carvings.

North of Salina in Ottawa County is a unique assemblage of giant rocks. These mammoth sandstone concretions are known as "Rock City". The incredible sight is 2.5 miles southwest of the town of Minneapolis. Having diameters of 20 feet and more, the weathered and rounded structures cover a large area which has been set aside as a park. Concretions of a more collectable size are found in McPherson County near Roxbury and just south of Freemount on State Highway 4.

Quartz-bearing gravels cap many hills in eastern Kansas. Among the minerals of these hard and resistant gravels are feldspar granite, basalt, quartz, agate, petrified wood, jasper and chert. Nuggets of native copper have also been found. Look in the uplands of the following counties for these rich glacial gravels: Anderson, Cowley, Elk, Greenwood, Lyon, Morris, Wabaunsee, Geary, Riley, Pottawatomie, and Marshall. The Big Blue River and its reservoirs in northeastern Kansas provide gem hunting areas. The banks and gravel pits near the Republican River also have quartz gems.

Lapidary materials along the Republican River north of Concordia

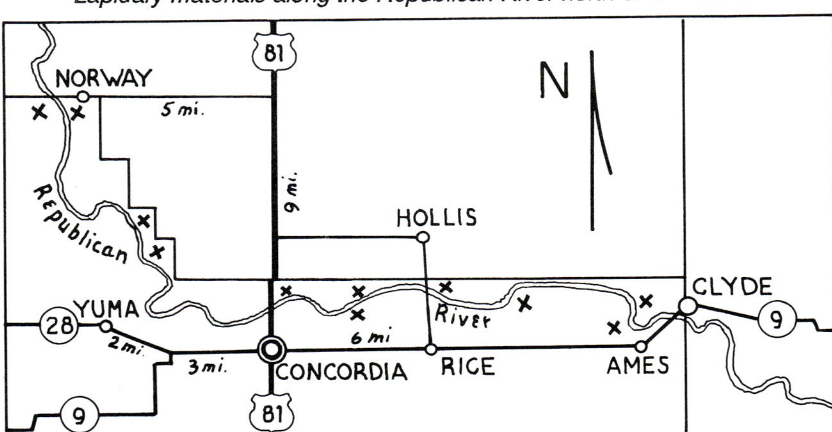

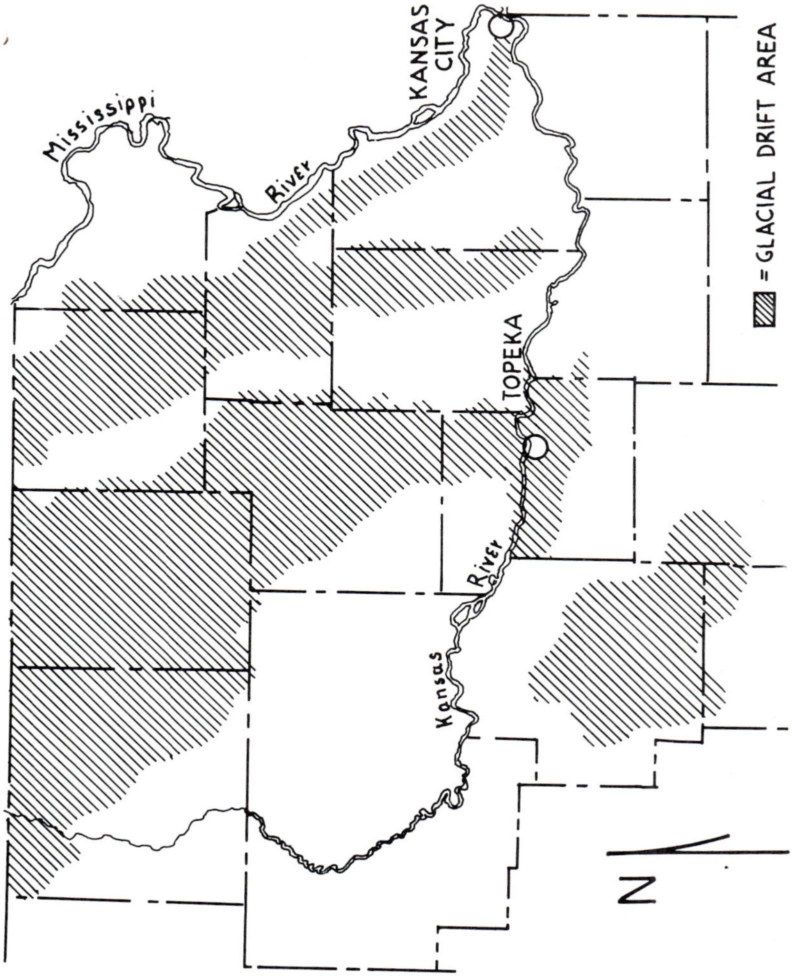

Quartz gems and fossils are found in glaciated areas of northeastern Kansas

Lake Superior agates are found in the Kansas City area and east of Topeka. Brachiopods, crinoids, cephalopods, belemnites and other fossils are abundant in the road cuts and quarries of the Kansas City area. Wyandotte County has several Pennsylvanian fossil outcrops near Bonner Springs.

Aragonite nodules are found in McPherson County 2 miles southeast of the city of McPherson. Dolomite crystals are present in the shales of this county. Geary County has many gravel quarries where petrified wood, agates, jasper, chalcedony and chert occur. Look also in road cuts and around dams for quartz, calcite and geodes.

Three eastern counties are known for their geodes. These are Chase,

Butler and Cowley Counties. The Cowley location is along the Walnut River, near a town appropriately named Rock. The Butler County location is north of Douglas, also on the Walnut River. The geodes are quartz, chalcedony and calcite. The Chase County locations are near Cedar Point and Chase City, along the Cottonwood River. Some cutting materials have been found near McLouth in eastern Jefferson County.

There are several items for the collector in Franklin County. Dolomite crystals are found in the Ross quarry of Williamsburg. In quarries near Ottowa aragonite and barite crystals are found in vugs and cavities in limestone. These same quarries are noted for intricately preserved plant fossils of the Pennsylvanian Era. The graceful seed ferns, scouring rushes and exotic trees flourished here in a mild, wet climate. An old quarry 4 miles south of Ottawa on U.S. 59 has yielded excellent specimens of club mosses, leaves, bark and seeds of this ancient forest. Other quarries are near Garnett.

Brown County has celestite crystals, pink in color, and sometimes in clusters shaped like rosettes. These unusual crystals occur northwest of Morrill 2 miles along Pony Creek. The crystals weather from the bluff east of the road on this private property. The most productive area has been badly abused in the past.

Pink celestite also occurs in southern Nebraska not far from the Kansas border. There are also geodes in Brown County. Agates, quartz gems, and petrified wood are found in the Pleistocene gravels south of Fairview. Algal limestone can be seen in road cuts and stream banks.

Woodson County has a native granite exposure. The cream-colored feldspar of this granite is accented by bluish quartz. The location is along U.S. Highway 75, 8 miles south of Yates Center. Drilling revealed peridotite beneath portions of the granite ridge. Quartz crystals have been found in this area, some of them pale amethyst in color.

Peridotite also occurs in Riley County in several places. It is a coarse-

Xs indicate pink celestite near Morrill, Kansas

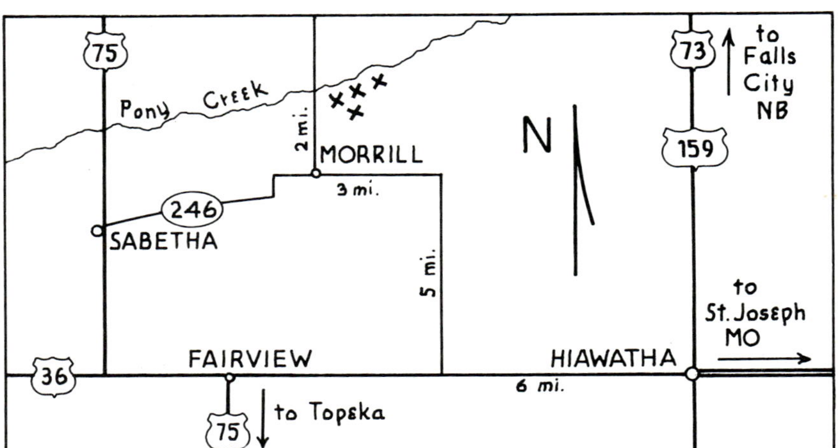

Calcite is one of the more common minerals of the Prairie States. This specimen is from Kansas.

graded basic igneous rock, gray-green in color and altering to yellow-gray clay with vermiculite. Five volcanic plugs are exposed at the surface in Riley County. The most important are near Bala and Stockdale. A low knob is 1 mile east and slightly south of Bala and a plug is located 3 miles east and 3 miles north of Bala. The Stockdale exposure is 4 miles west and 3 miles north of Stockdale in a stream bed. The peridotite contains ilmenite and garnet, and it is being investigated for the possibility of diamonds.

The Verdigris River in Greenwood, Wilson, and Montgomery Counties has yielded freshwater pearls. Agates and other gems have been found along the river. Fossil jellyfish were found in Greenwood County and numerous fossils occur near Fredonia in Wilson County.

Near the southeast corner of Kansas is Labette County where there is a fossil reef 12 miles long and 13 feet deep. Among the numerous fossils found here are corals, brachiopods, clams, oysters, and bryzoans. There are also crinoid stems and fusulinids. The fossil shells occur in the reef limestone in small concentrations.

Cherokee, the corner county, is of exceptional interest. It is part of the Tri-State district, one of the world's legendary zinc-producing areas.

The lead-zinc mines of this famous district have produced thousands of exquisite museum and cabinet specimens of showy crystals of many species. Some of the combination specimens are superb and dramatic in appearance, and are widely sought by selective collectors. Among the choice specimens are smithsonite, sphalerite, galena, barite, marcasite, pyrite, quartz, calcite, dolomite, hemimorphite, cerussite, chalcopyrite, and more. Large transparent yellow calcite crystals are popular collector's items, as well as spectacular specimens of shining cubes of galena and cockscomb marcasite. Ruby sphalerite with chalcopyrite, dolomite or calcite is colorful and resplendent. The old mines are closed now. For many years club field trips were allowed with experienced guides, and fine specimens were found each time. Rock shops and private collectors in the area sometimes have specimens for sale or trade.

The dominant features in this area are the "chat" piles from the bonanza days of the mines. Chat is a corruption of the word chert. Columbus and Baxter Springs residents may be able to help collectors locate specimens.

The mineral notes of Kansas would not be complete without mentioning the extraordinary number of meteorites which have been found in the state. Meteorites have been reported from 50 counties. Part of the reason for the unusual number of recoveries of these outer space visitors dates back to the "meteorite farm" of 1885 which made the entire state meteorite conscious.

The famous farm, between Greensburg and Haviland in Kiowa County, had yielded thousands of pounds of meteorites. Many were found while a well was being dug. The state-wide search for meteorites has been aided by widespread plowing and the lack of thick vegetation. The predominance of sedimentary rocks has also made it easier to recognize meteorites. In fields used for agriculture there are seldom rocks on the surface of the ground, so every rock is examined with curiosity.

The tons of meteorites from the crater southwest of Haviland are displayed in museums around the world, and are being carefully analyzed by space scientists. A ton size specimen is on exhibit in Greensburg. All of the specimens here are classed as Dallasites, or stony irons. During the dust bowl days, farmers found many meteorites in arid fields in this county.

Other meteorites have come from Pratt and Iuka in Pratt County which is east of Kiowa County. Several finds are listed for each of the following counties: Sherman, Norton, Smith, Greeley, Graham, Osborne and Sumner. There is a possibility of finding a meteorite in any county in Kansas, or for that matter anywhere in the prairie states. Every collector should be aware of that possibility and should know some of the characteristics of meteorites in order to be able to recognize them.

A primary aspect of meteorites is that they appear burned, or as if they had

Cordities, a fossil plant can be found in Pennsylvanian outcrops.

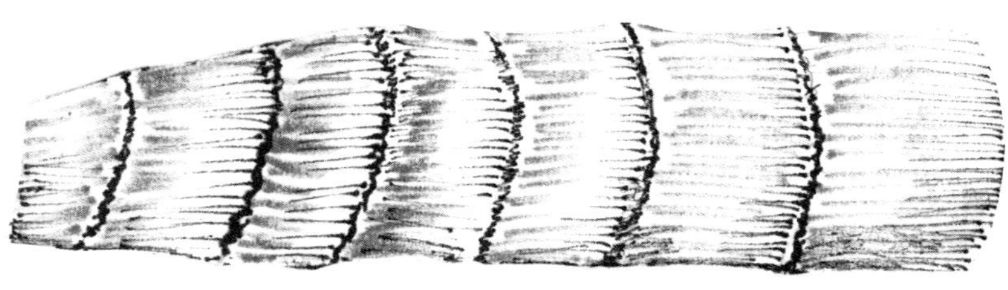

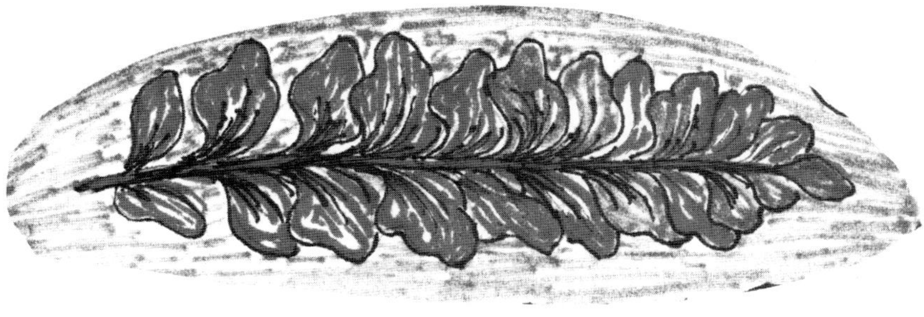

Neuropteris is one of the Pennsylvanian fossils occurring near Ottawa.

been partially melted. They are generally pockmarked and pitted. They may look like a weathered piece of iron. Meteorites can be solid metal, stone or a mixture of both. They are not porous and they do not have vugs or cavities lined with crystals.

Most meteorites, because of their iron content, will attract a magnet. Iron meteorites are easier to identify than "stonys". Iron and nickel can usually be seen, and the iron meteorites are usually much heavier than any of the surrounding materials.

Stony meteorites consist of heavy igneous minerals and also contain metallic iron, however since they superficially resemble certain other igneous rocks, they are more difficult to identify with certainty. It often takes a mineralogist connected with a university or museum to authenticate the finds.

The surfaces of many meteorites show streamlined channels produced by the hot gases rushing backwards from the tip of the object as it flew towards earth.

A meteorite is most often quite different from the country rock. It does not splinter easily. A meteorite does not look like slag, lava, or a clinker. Many meteorites which have lain on the ground for a long time have a brown crust.

A suspected meteorite should be ground on the end with a lapidary grinding wheel until the surface is smooth and even. When examined with a magnifying lens, silvery grains of nickel will be seen. Test the grains with a needle to be sure they are not mica or pyrite. A meteorite is a rare and valuable find.

In spite of the fact that Kansas gem, mineral and fossil locations are privately owned, the many hobbyists of the state are still finding places where they can get permission to collect an interesting variety of material. Clubs have productive field trips, and Kansas lapidaries have become expert at original work.

There are excellent rock shops in many of the major cities of Kansas and some of the smaller ones as well. Some of these are located in Emporia, Hill City, Manhattan, Mulvane, Overland Park, Topeka and Wichita.

The greater Kansas City Show every spring has many displays of native fossils, rocks and gems, and exhibits of the fine craftsmanship of the lapidaries of the Kansas and Missouri areas.

Kansas has earth science collections at Hays, Lawrence, Lindsborg, Emporia, McPherson, Topeka, Baldwin City, and Salina. The fossil collection at Hays is particularly noteworthy.

Rock City in northern Kansas is a famous site of immense concretions. Photo courtesy of Kansas Geolgocial Survey.

A number of important fossil discoveries have been made near Castle Rock, a 70 foot chalk spire south of Quinter. Photo by Kansas Department of Economics.

Xs indicate primary collecting counties in Missouri

MISSOURI

Missouri is a scenic state of great diversity. There are two great rivers, the Mississippi and the Missouri, large lakes, Lake of the Ozarks and Table Rock, plus mountains and forests, fertile valleys, and two major metropolitan areas, St. Louis and Kansas City.

A mecca for fossil collectors, Missouri also has major drawing cards for mineral collectors, for example barite and geodes, and several fine areas for collectors of lapidary materials, lace agate, mozarkite, banded agate, petrified wood and coral.

There are numerous well-placed state parks, and also National Recreation areas in the Mark Twain and Clark National Forests. Outside of these all of the land is privately owned. At present many areas are posted. It is possible to get permission to collect in most of the better-known regions, and there are also fee areas which give the collector a good chance to pursue his hobby.

Western and Central Missouri

Western Missouri gravels, along the Missouri River and its tributaries, sometimes yield petrified wood, jasper, carnelian and agate, Some of the agates are choice examples of the popular Lake Superior agates. Large and colorful "Lakers" have been found along the Nodaway and Platte Rivers and the streams which feed them in Nodaway and Andrew Counties in northwest Missouri. Both the Nodaway and the Platte flow into the Missouri. Look along the banks, in sand bars, excavations and gravel operations wherever you can get permission.

Pennsylvanian fossils are found in the strip coal mines at Bronaugh in southern Vernon County, Rich Hill in Bates County, Appleton City in St. Clair County, and northwest of Clinton in Henry County. Marcasite and pyrite crystals are also found in these mines.

The most significant mineralized area of western Missouri is the lead-zinc "Tri-State" district near Joplin, long the source of some of the most esthetic crystals in the world formed from lead and zinc minerals and associated minerals. At present the old and dangerous mines are closed to visitors.

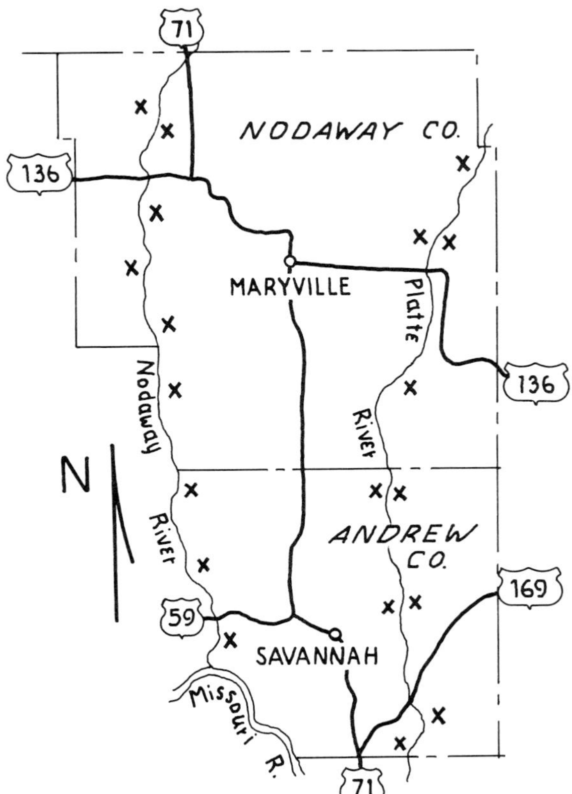

Agate locations in northwest Missouri

Brown's Minerals of Joplin reported that there are sometimes dumps near Joplin open for collecting, but that minerals there were not crystals, but study type samples. Brown's wrote that visitors should stop to see the Mineral Museum of the city of Joplin in Schefferdecker Park. A showy selection of crystals from the famous "Tri-State" mines is on display. There is no admission charge.

Many species of crinoids come from the Burlington limestone of Boone and Callaway Counties and other adjacent counties in the central part of the state. Fossil exposures are north of Williamsburg along Cow Creek, and east of New Bloomfield. Excellent specimens of marcasite and pyrite crystals have come from the strip coal mines of Boone and Randolph Counties.

Platte, Buchanan, Andrew, Clay, Clinton and DeKalb Counties have abundant brachiopods, pelecypods, and gastropods, representing many species. Look for them in Pennsylvanian exposures in quarries, stream banks and road cuts.

Missouri's most popular gem material is in Benton County, near the town of Lincoln. It is the State Rock "mozarkite", a colorful "agatized" chert. The brightly patterned material is found in-place in Hickory County, Dade County, and adjoining counties, and as float in many more places. Named for the Missouri Ozarks, the chert is fine-grained, and grades into chalcedony and agate. Drusy quartz crystals enhance some specimens.

Colors which predominate are red, pink, rose, yellow-orange, plum, maroon, and blue-gray. Although the usual patterns are swirly "cloud-like" or "flame-like" forms. the imaginative and lucky collector can spot all sorts of pictures suitable for scenic or picture cabochons.

Mozarkite occurs as nodules, geodes, and masses in road cuts, stream beds, and banks and ditches around dams and lakes in several regions centered by Lincoln. Large amounts of material were first revealed in road cuts and during construction of an artificial lake in Benton County. The chert boulders weather out of dolomitic limestone.

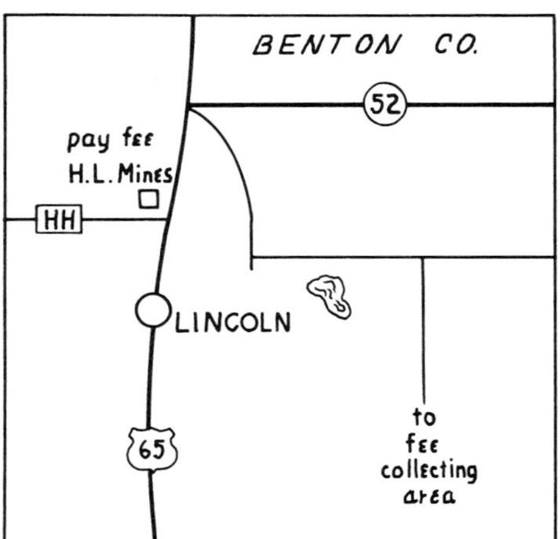

Fee mozarkite area near Lincoln

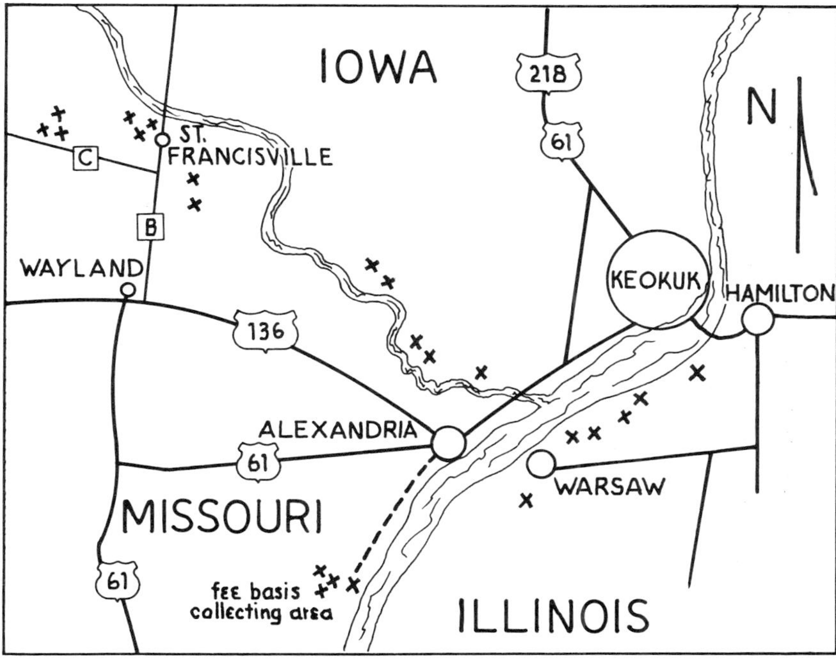

Geodes near Alexandria and St. Francisville

Although pieces of colorful chert can still be found in the gullies and ditches along the roads near Lincoln, the best picking is in a fee basis area east and north of Lincoln. Mr. Lineville Harms who lives in Sedalia, is the operator of the fee area. The manager for the collecting site lives 3 miles north of Lincoln just northwest of the junction of U.S. 65 and County road "HH". The fee is modest and there is no limit.

Barite is also found in Benton County, south of Warsaw and east of Lincoln. Other Minerals of Benton County are coal, limonite and hematite. Barite crystals have been found in Morgan, Moniteau and Cole Counties. Clays have been mined in this entire area.

Montana-type dendritic agate and Lake Superior banded agates as well as petrified wood and other quartz gems are found along the Missouri River in the gravel bars on either side of Columbia.

Eastern Missouri

Spectacular geodes of the Warsaw formation are found in Clark County. Although the geodes are found in stream beds and banks, in road cuts and excavations, and in fields, a prime spot is a fee area known as Scheffler's Geode Mine. Scheffler's rock shop is at the junction of Routes 136 and 61. The mine is located 8 miles south of Alexandria just off U.S. 61. St. Francisville is also famous for geodes.

The quartz geodes of northern MIssouri often have inclusions of bright pink calcite. Some geodes sparkle with pyrite crystals. Others have iridescent calcite, banded chalcedony, smoky quartz, hematite, galena, goethite, limonite, millerite, common opal and a host of other inclusions. An especially beautiful variety has "dew drop" crystals of limpid quartz.

Gravel quarries along the Mississippi River have been rich in quartz gems for generations of collectors. There are Lake Superior agates, jasper, chalcedony, quartz nodules of several colors, petrified wood, carnelian and other choice lapidary materials. The noted La Grange quarry is reported closed, but there are other gravel operations between Keokuk and St. Louis where collecting is sometimes allowed. Gravel piles are to be found in Pike and Lincoln Counties. Limestone quarries are operated in Lewis, Ralls and Pike Counties.

Missouri Lace Agate, banded agate, drusy quartz, and barite are popular collectables from Washington County. The strip mines are in the vicinity of Washington State Park. Mines are located near the park and south to Old

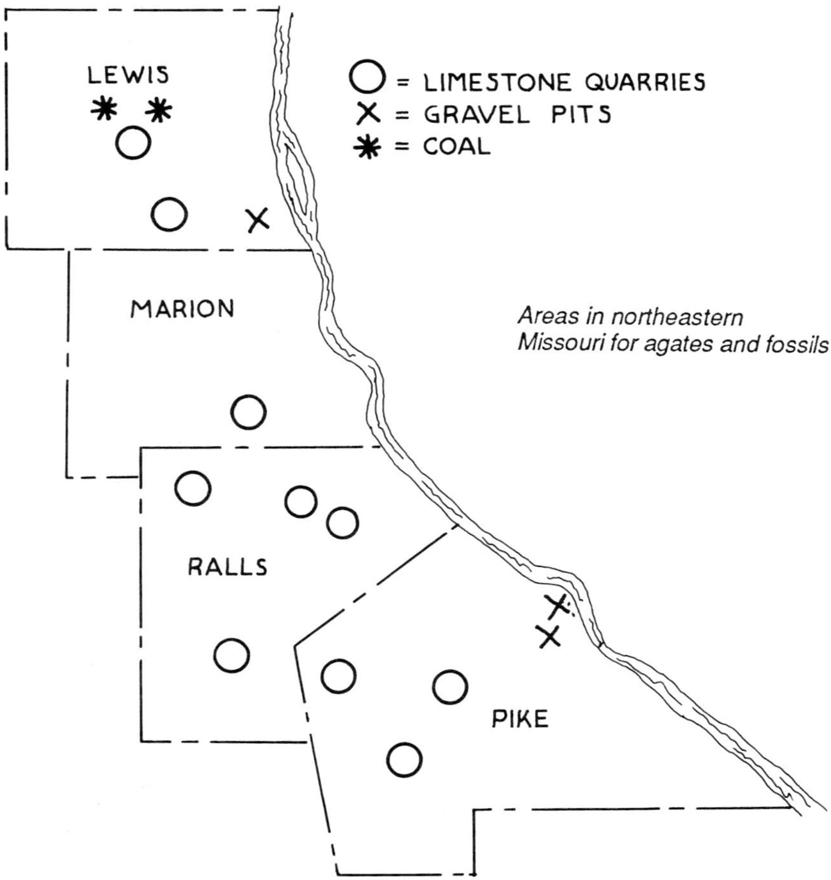

Areas in northeastern Missouri for agates and fossils

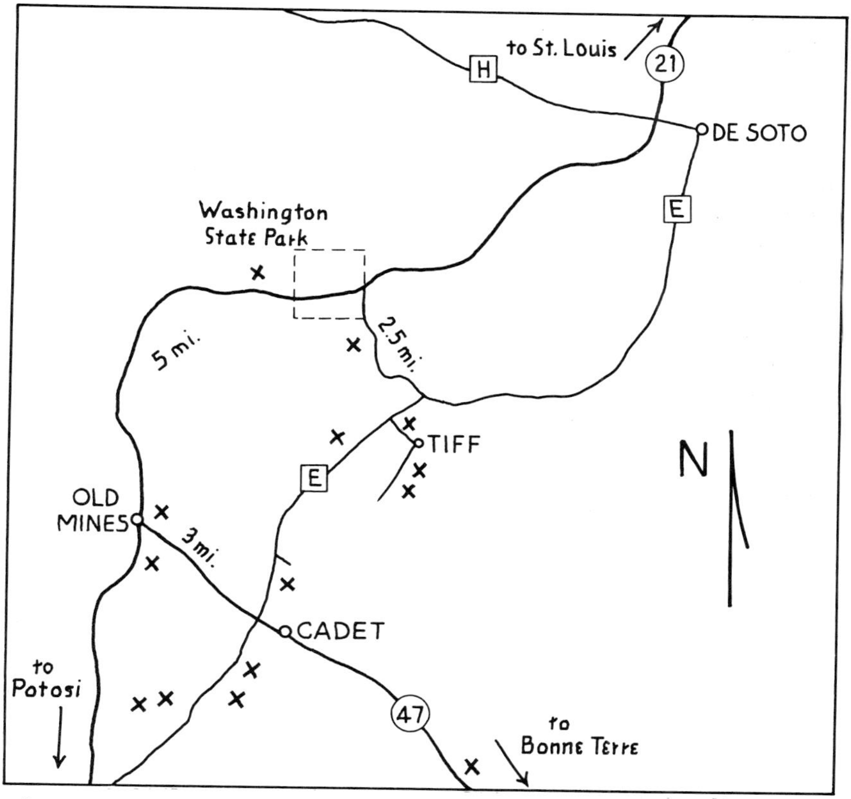

Access routes to lace agate collecting at mine dumps near Washington State Park

Missouri has lace agate in several locations still open to collectors near Washington State Park. Photo by Willis Smith.

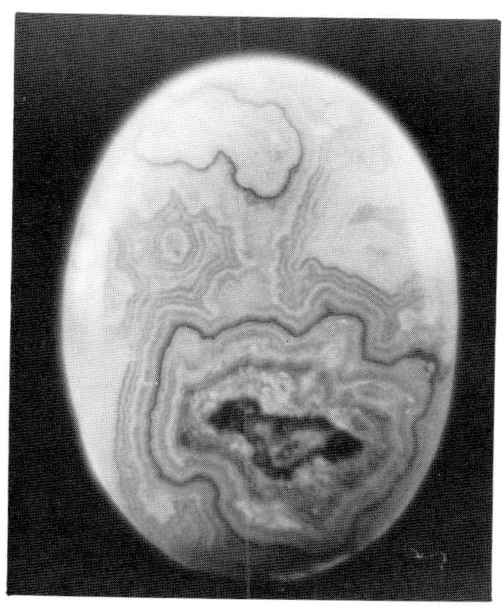

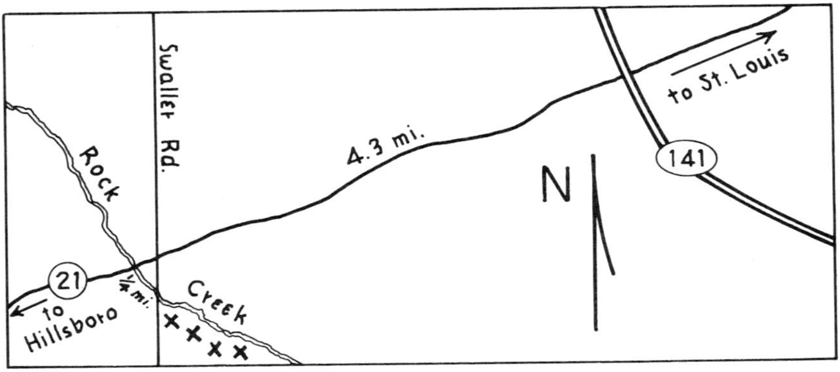

Cutting material south of St. Louis

Mines, east to Cadet, and north to De Soto. A favorite area is at Tiff, 1 mile off Highway "E" and 5 miles east of Washington State Park. Another area is near Potosi. The Missouri Lace agate is a seam agate in shades and tints of blue, gray, white, and sometimes red and yellow. Well-selected agate takes an exceptional polish and makes distinctive cabochons.

The St. Louis area has furnished many an exciting field trip for local rock hobbyists. Road cuts near St. Louis provided surprised and delighted collectors some years ago with many vibrantly colored fortification agates. Called Union Road agates, these brilliant patterned nodules were encased in white chert. Most of the first collecting sites are buried beneath the highways, but new construction frequently reveals more of the prized agates to persistent collectors. As late as the summer of 1987, Willis Smith was able to find beautiful agates in new excavations. Similar nodules come from the south side of the Meremac River near Arnold. Colorful chert is found on Rock Creek south of the intersection of Highways 21 and 141.

Three miles south of the Barnhart exit from I-55, south of St. Louis, outcrops along the service road parallel to the Highway, have numerous fossils such as brachiopods, gastropods, and pelecypods. One of the materials here resembles the so-called "turritella" agate. It is a dolomitic limestone, with a marble-like texture, which is polishable.

Gravel operations in the vicinity of St. Louis have produced the usual quartz family cutting materials, often of exceptional quality. Colorful quartz gems are also found along the lower stretches of the Missouri River prior to its confluence with the Mississippi. It is best to contact local clubs before doing any collecting in the St. Louis area, according to Willis Smith, who is a well-known collector, lapidary, and member of the McDonnell Douglas Gem and Mineral Club.

St. Genevieve County has silicified corals, sponges, blastoids and other fossils along Little Saline Creek. Marine fossils are also numerous along Establishment Creek. The quarries of the Ozora Marble Company have many species of brachiopods as well as gastropods, pelecypods, and more.

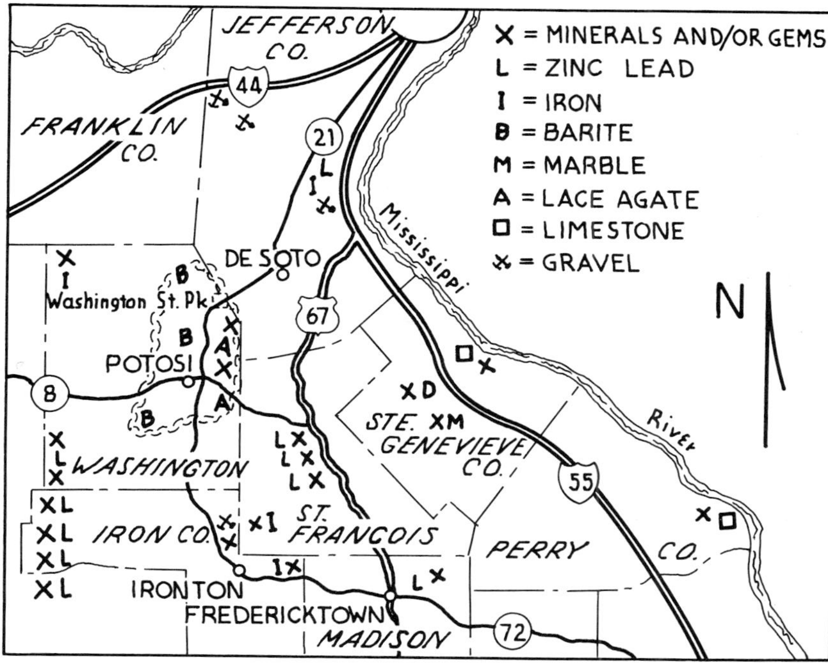

East central Missouri mining and collecting areas

An area with plentiful cutting material is at Farmington in St. Francois County west of Highway 67 and north of County Road "W". Agates, drusy quartz, and colorful chert are found near the power lines.

Pilot Knob, north of Ironton, in Iron County, is a source of intense red jasper. Iron County is noted for more minerals than gems. The important Viburnum Trend lead-zinc mines cross the county in the northeast part.

Agates and chert near Farmington

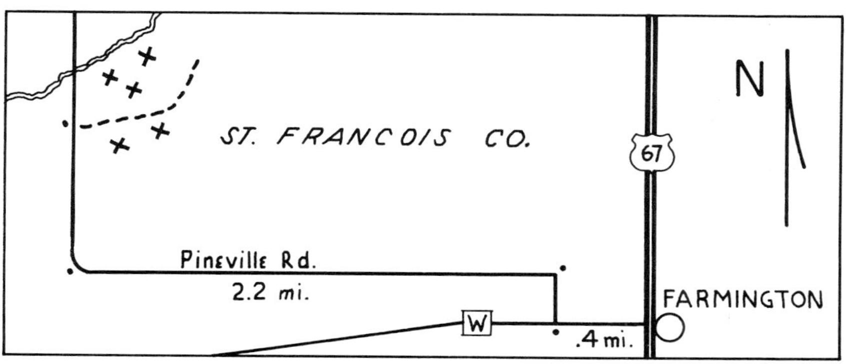

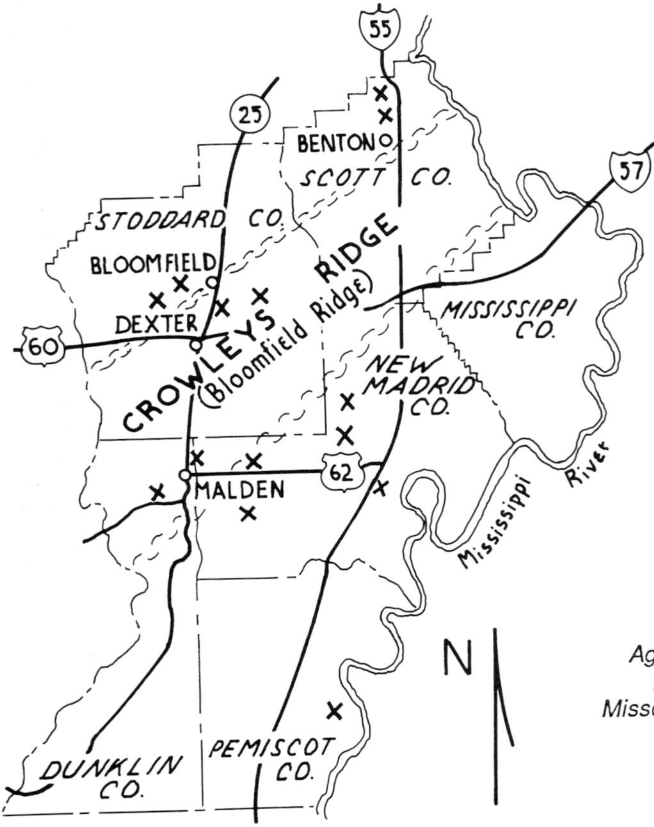

Agates and cutting materials in Missouri's "bootheel"

Copper, iron, and magnetite come from this county. There are several areas of the Clark National Forest in the St. Francois Mountains which cover parts of several counties in southeast Missouri. This is an igneous uplift. West of Fredericktown in Madison County, the Forest Service has a campground near the Einstein Silver Mine . The dumps are said to reward ambitious collectors with galena, fluorite, sphalerite, and mica.

In the "bootheel" counties of southeastern Missouri there is an important gravel producing area known as Crowley's Ridge or Bloomfield Ridge. Deposits are being worked in the uplands of this ridge in Scott, Stoddard, and northern Dunklin Counties. Other gravel operations in Missouri's "bootheel" are in the channel and floodplain gravels of Mississippi, New Madrid and Pemiscot Counties. The Tertiary gravels of Crowley's Ridge attract gem material collectors and fossil collectors. Among the cutting materials are Missouri Lace, black chert, brown, red and yellow jasper, chalcedony, carnelian, jasp-agate, silicified coral, and petrified wood. Fossils, in addition to lapidary varieties, include shark's teeth and other marine species.

The ridge extends from southwest to northeast from near the town of Malden near the Dunklin-New Madrid County Line to the town of Anna, across

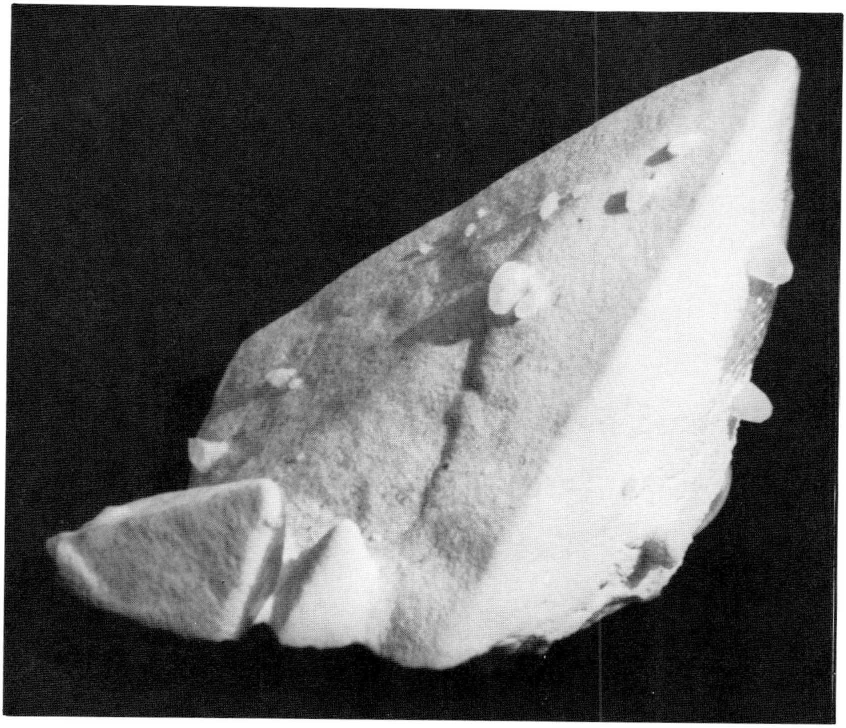

This frosted calcite crystal is from southwest Missouri.

the Mississippi River in Illinois. The "bootheel" counties are drained by the Mississippi and the St. Francis Rivers. There are several gravel operations in Stoddard County near Bloomfield, with others located near Benton in Scott County, Marston in New Madrid County and Caruthersville in Pemiscot County. An especially good area is 6 miles west of Malden near the county line of Dunklin and New Madrid. Willis Smith of the McDonnell Douglas Club says that there are some productive pits west and south of Dexter in Stoddard County. He warns that pit operators change frequently, so visitors should always find out which places are open to collecting.

Shimmering fresh water pearls have come from several Missouri rivers, especially the Current River and the Meremac River. A productive area was near Sullivan in Crawford County.

Missouri is a land of many caves. More than 20 caves are open to the public. Experienced cavers have explored many others. Some private caves may allow exploring by qualified cavers. McDonald County has four caves, and Stone and Camden Counties each have three. Others are scattered in hilly regions. There has been abundant ground water in the Ozark region, so some of these incredible caves are classed as "live". That is, enough water is present in the caves that stalactites and stalagmites are still forming. Delicate

Built like stairsteps, this Missouri calcite has inclusions of chalcopyrite

and fragile crystals of snowy aragonite and honey-colored calcite also occur in these caves. When first explored some of the caves contained arrowheads and other evidences of early man. Pleistocene vertebrate fossils were found in many caves, among them Bat Cave, Carroll Cave, Cox Cave, Nameless Cave, and Stark Caverns.

Galena is the State Mineral of the "Show Me" state. This specimen comes from the Tri-State Region near Joplin. Zeitner Collection.

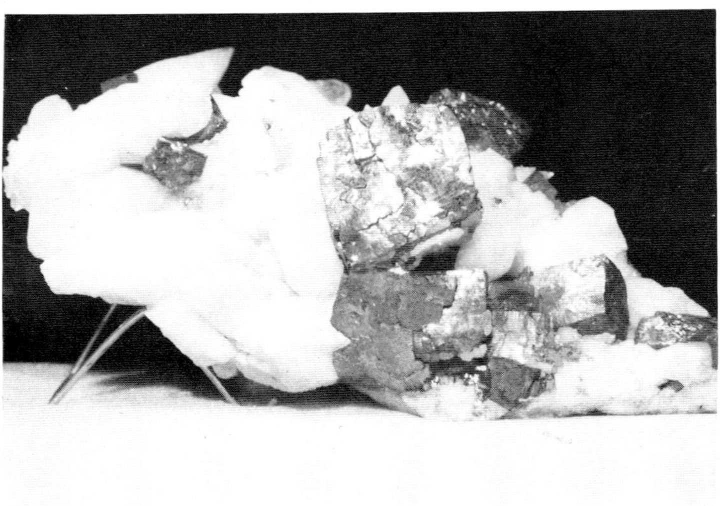

Caving in undeveloped caves is extremely dangerous, and should not be undertaken without experienced and qualified leaders. Even trained spelunkers often run into unexpected problems.

Several of the caves are Federal and State Parklands. These are Ozark Caverns at Osage Beach, Fishers Cave at Sullivan and Round Spring Cave at Round Spring. The Meremac Caverns at Stanton and Mark Twain Cave at Hannibal are also well-known. Other caves of note are the Bride Cave of Camdenton, Crystal Cave of Springfield and Onyx Cave of Van Buren.

The Missouri Geological Survey offers some advice for cave explorers:

1. Never enter a cave alone.
2. Your party should be properly equipped and led by someone experienced, and preferably familiar with the cave.
3. Wear suitable clothing including a hard hat.
4. Have three sources of light: carbide light, flashlights, and candles.
5. If there are signs of a large stream near the cave or if there has been recent flooding in the area, stay out.
6. Pit- or sinkhole-type caves are the most dangerous. One needs thorough knowledge of climbing techniques, rappelling, belaying, and prussik climbing and an experienced belayer before entering such a cave.
7. Before exploring, contact the Missouri Speleological Society, in care of the Missouri Geological Survey at Rolla.
8. Do not destroy cave formations.

Vertebrate fossils are found in several places in Missouri in addition to the caves mentioned. The best area for Pleistocene vertebrates is in the center of the state in Howard, Boone, Cooper and Moniteau Counties.

Numerous vertebrate bones and teeth were found 1.25 miles northwest of Kimmswick in a swampy area along Rock Creek, Jefferson County. Another location is 1 mile west of Herculaneum on Joachim Creek. Among the fossils have been mastodons, horses, musk oxen, sloth, bison, moose and deer. In Pettis County 4.5 miles from Sedalia, swampy land on Flat Creek yielded mastodon bones. In Lewis county vertebrate fossils were found in the gravels 2 miles south of La Grange. A large sink hole called Sullens Mine had many fossils: sloth, turtles, tapirs, mastodons, and more. It is in Moniteau County 1 mile southwest of Enor.

Vertebrate fossils must be collected with extreme care. Read how to collect them properly and how to preserve them in any basic paleontology text. Be sure and keep an accurate record of exactly where the fossil was found, including such details as whether it was found in a stream or in a road cut, or an eroding bank. If the fossil is large, unusual, or if it appears there are many more bones there, contact a professional paleontologist at the nearest university or museum.

There is a wealth of material in Missouri for the patient and informed collector. Most of the sites are privately owned, so permission is necessary. Since several of Missouri's most exotic materials have been fairly recent discoveries, it seems likely that increased prospecting will result in additional discoveries of gems and fossils.

The fee basis areas in Missouri are in the geode area near Alexandria, and sometimes St. Francis, in the Mozarkite area near Lincoln, and near Drury at Mr. Magruder's "Spring Valley Mine".

There are earth science collections in Missouri at colleges or libraries in Jefferson City, Columbia, Canton, St. Louis, Fayette, Fulton, Kansas City, St. Joseph, and Rolla.

There are numerous rock shops in the state. Several are in the St. Louis and Kansas City metropolitan areas. Others are in Springfield, Alexandria, Carthage, Jefferson City, Joplin, Raytown and Lincoln.

Missouri has several outstanding shows. The McDonnell Douglas Club show and the Rock Hobby Club show and others in the St. Louis area, and the Kansas City show are examples. Some of the collectors of gems, minerals, and artifacts are very advanced, and the displays you see at these shows are of professional quality. There are some outstanding clubs.

Visiting collectors will find that Missouri hobbyists are friendly, enthusiastic, and glad to be of help.

Mozarkite is a multicolored, brightly patterned chert found in several Missouri counties.

Follow the waterways for fossils and cutting materials. Photo by Walker, Missouri Tourism.

SUMMARY

These states are all becoming more conscious of the boost to the economy that rockhounds bring. Rockhounds are travellers, outdoors people who are enthusiastic about beautiful scenery and natural wonders. If we find some good specimens while on a vacation we stay longer, spend more, and a return trip is assured. Bureaus of Tourism and Chambers of Commerce are beginning to realize this. Write ahead for information, and if you get it and make the trip be sure to write and tell them about it. The more these agencies know about the rock collecting hobby, the more they will be able to help us.

If you are club members be sure to publicize your shows and swaps in the states you hope to visit and ask that they send you information on their special events.

These six states of the Midwest have produced great amounts of gold, lead, zinc, coal, oil, quartz, feldspar, sandstone, limestone, sand, gravel, and happily for us many thousands of exquisite crystals, colorful agates, and fascinating fossils. As the development of the states continues, new locations will be found for known minerals and fossils, and new species of minerals and fossils will be discovered.

We can help the sciences of mineralogy and paleontology by sharing information and finds. It is important to our hobby to help scientists write the history of our earth.

Our hobby is also influential in the arts. Skilled lapidaries are producing elegant jewelry, graceful carvings, and shining art objects of durable stone. We should encourage the teaching of the lapidary arts in schools, along with the other fine arts.

As members of a unique hobby which combines science and art we have much to offer our communities, so let your finds from these states add to knowledge and beauty.

BIBLIOGRAPHY

General

Bailey, R.H., GLACIER, Time-Life, Alexandria, VA 1982

Fenton and Fenton, THE FOSSIL BOOK, Doubleday, New York 1958

MacFall, Russell, GEM HUNTERS GUIDE, Crowell, New York 1978

MacFall, Russell and Wollin, Jay, FOSSILS FOR AMATEURS, Van Nostrand, New York, 1972

MacFall, Russell, ROCK HUNTERS GUIDE, Crowell, New York 1980

Sinkankas, John, GEMSTONES OF NORTH AMERICA, Van Nostrand 1976

South Dakota

Petch, Bruno and McGregor, Duncan, SOUTH DAKOTA'S ROCK HISTORY, South Dakota Geological Survey, Vermillion 1976

Petch, Bruno and McGregor, Duncan, MINERALS AND ROCKS OF SOUTH DAKOTA, South Dakota Geological Survey, 1973

Roberts, Willard and Rapp, George, MINERALOGY OF THE BLACK HILLS, South Dakota School of Mines, Rapid City 1965

Nebraska

Boellstorff, John, "Collecting in Nebraska's Glacial Deposit", *Rocks and Minerals*, Volume 55 number 3

Pabian, Roger, MINERALS AND GEMSTONES OF NEBRASKA, Nebraska Geological Survey, Lincoln 1971

Pabian, Roger and De Graw, Harold, "Collecting in Nebraska's Cretaceous Strata", *Rocks and Minerals*, Volume 55 number 3

Pabian, Roger, "Collecting in Nebraska's Pennsylvanian-Permian Strata", *Rocks and Minerals*, Volume 55, number 3.

Pabian, Roger, RECORD IN ROCK, A HANDBOOK OF NEBRASKA'S INVERTEBRATE FOSSILS, Nebraska Geological Survey, Lincoln 1970

Kansas

Tolsted, Laura and Swinford, Ada, KANSAS ROCKS AND MINERALS, Kansas Geological Survey, Lawrence 1979

Williams, Roger et al, ANCIENT LIFE FOUND IN KANSAS ROCKS, Kansas Geological Survey, Lawrence 1975

Buchananan, Rex, KANSAS ROCKS AND MINERALS, Geological Survey, Lawrence 1986

Missouri

Keller, W.D., COMMON ROCKS AND MINERALS OF MISSOURI, University of Missouri Press, Columbia 1969

Mehl, M.G., MISSOURI'S ICE AGATE FOSSILS, Missouri Geological Survey, Rolla 1962

Unklesbay, A.G., COMMON FOSSILS OF MISSOURI, University of Missouri Press, Columbia 1969.

Iowa

Horick, Paul et al, THE MINERALS OF IOWA, Geological Survey, Iowa City 1974

Lemish, J., MINERAL DEPOSITS OF IOWA, Iowa Southern Utilities Co., Centerville 1969

Rose, J.N., FOSSILS AND ROCKS OF EASTERN IOWA, Iowa Geological Survey, Iowa City 1967

APPENDIX

SOME HELPS FOR FIELD IDENTIFICATION

The Mohs Scale of Hardness

1 Talc
2 Gypsum
3 Calcite
4 Fluorite
5 Apatite
6 Feldspar
7 Quartz
8 Topaz
9 Corundum
10 Diamond

Minerals of these states grouped by luster

Metallic

galena, gold, pyrite, chalcopyrite, marcasite, millerite

Sub-metallic

hematite, goethite, pyrolusite, columbite, tantalite, uraninite, cassiterite sphalerite, rutile

Vitreous (glassy)

black or gray - quartz, tourmaline
red-brown & brown - garnet, staurolite, calcite, siderite, barite
orange - calcite, quartz, beryl
yellow - quartz, opal, spodumene, fluorite, aragonite, barite
red - tourmaline, calcite (pink) quartz (rose) orthoclase, (pink) microcline (peach pink) almandite
green - vivianite, apatite, beryl, tourmaline, hornblende, spodumene, epidote, grossularite
blue - beryl, celestite, vivianite, apatite
violet - quartz, calcite, fluorite, spodumene almandite
colorless and white - quartz, gypsum, calcite, opal, orthoclase, microcline, dolomite, beryl

Pearly

gypsum, muscovite, biotite, lepidolite, phlogopite

Silky

black - goethite, hematite
white - gypsum, aragonite, calcite
brown - siderite, barite, calcite
pink - asterated rose quartz, gypsum

Earthy

black - pyrolusite, manganite, goethite
white - clay
brown - limonite
yellow - carnotite
red - hematite
green - chlorite

Crystal habits of minerals in this book

ISOMETRIC

cube - pyrite, galena
octahedron - fluorite
dodecahedron - garnets

TETRAGONAL

hexagonal prism - apatite, calcite, quartz, beryl
hexagonal tabular - hematite apatite, beryl
rhombohedron - hematite, calcite, dolomite, siderite

ORTHORHOMBIC

prismatic - aragonite, cerussite, barite, celestite
dipyramidal prismatic - staurolite
tabular - marcasite, columbite, tantalite, barite, celestite
bladed - goethite

MONOCLINIC

prismatic - manganite, gypsum, spodumene
wedge shape - gypsum
pseudohexagonal tabular - muscovite, biotite, lepidolite, chlorite, phlogopite

TRICLINIC

blocky - microcline
bladed - albite